Life science - concepts analysed in a different way

Volume 1

Darani vasudevan

About the contents inside.....

The contents of the book are prepared in such a way, that it will make the readers to analyse the topics we usually read in any standard or preferred textbooks in a different manner like comparing the theoretical contents with life events or any research observations. The idea behind this, is the questions asked in the part C of CSIR UGC NET examination conducted widely throughout India for the post of Assistant Professor and Junior research fellow. This book is filled with concepts regarding various subdivisions of Life Science like Systematic, ecology, biochemistry, cell biology, genetics, molecular biology, immunology, physiology and evolution. Most of us don't have the practise of analysing or comparing the idea behind the results we obtain in our research and projects with the textbook contents. The main idea of Part C is to make us relate both theory and practical. This book will help the readers to prepare for Part C in a better way than to simply read text books that is not fruitful idea to succeed in that section of CSIR UGC NET examination. Hope you all enjoy and find it useful!!!!

V.Darani M.Sc.,M.Phil.,SET

CONCEPTS RELATED TO CELL BIOLOGY, GENETICS AND MOLECULAR BIOLOGY

CELL DIVISION

There are three types of cell divisions. They are,

- Amitosis
- Mitosis
- Meiosis

The summary of events occurring in each of these stages are as follows:

Amitosis:

In this simple cell division, the nucleus in the cell elongates and a constriction appear in the centre. This constriction gradually gets deeper giving rise to two daughter nuclei. The two daughter nuclei thus formed are not equal in size. This nuclear division is followed by constriction of the cytoplasm. If constriction of cytoplasm doesn't occur it results in the formation of multinucleate cells named Coenocyte (plants) and Syncytium (Animals).

Mitosis:

(Somatic – Homotypic – Equitorial division)

The mitosis was first demonstrated by W. Flemming in the year 1882. In the case of mitosis the daughter cells formed are quantitatively and qualitatively alike like parents. It includes two parts karyogenesis and cytokinesis. Healing of wounds and damaged organs are a part of mitotic division. It includes five stages based on change in morphology of nucleus and chromosomes. The stages are

- Interphase
- Prophase
- Metaphase
- Anaphase
- Telophase

Interphase:

i. It is the non dividing stage

ii. Chromosomes are invisible in light microscope
iii. Chromatin is present in granular or thread forms
iv. It is the resting stage during which many physiological and biochemical changes occur in the cell.
v. The biochemical reactions include DNA replication, protein and RNA synthesis.
vi. In a cell cycle of 24 hours 23 hours are spent in interphase and the mitosis occurs in a duration of one hour.
vii. Interphase is divided into three phases named G_1 phase, S-phase and G_2 phase.
viii. G_1 phase- This is the gap in the cell cycle during which protein and RNA synthesis occurs. It consumes 30-40% of the total time in interphase.
ix. S phase: During this phase synthesis and replication of DNA results in the formation of sister chromatids. All histones of the nucleus are utilized during this phase. At the end of S phase, each chromosome is composed of two genetically and morphologically identical sister chromatids.
x. G_2 phase: It is the second gap in the interphase but its duration is short. This phase is also marked by RNA and protein synthesis. Some proteins prepared during this phase are necessary for entering the mitotic phase.

Prophase
i. Chromosomes appear as fine threads.
ii. Nucleus appear as ball of wool.
iii. Chromosome become thicker and thicker due to continuous dehydration and removal of water, coiling and contraction.
iv. Each chromatid possess one complete set of DNA (i.e., two chromatids attached by a centromere)
v. The disappearance of nuclear membrane and nuclo;us mark the end of prophase.
vi. During late prophase, all the thickened and shortened chromosomes are visible lying near the centre.

Metaphase:

i. Spindle fibres starts to appear and get attached to chromosomes at the point of centromere.
ii. Chromosomes are visible lying on the equatorial and metaphase plate (metaphase plate is and imaginary line for understanding the concept of cell division).
iii. Chromosomes are free floating in the cytoplasm.
iv. Spindles are made up of tubulin which is a protein and is a part of microtubule.
v. The characteristic features are absence of relational coiling between daughter chromosomes (i.e) the two sister chromatids are coiled in relation to each other

Anaphase:
i. It is the migratory phase.
ii. Each centromere divides transversely to separate the two sister chromatids which move to different poles of the cell.
iii. The chromosome movement is due to shortening and shrinking of spindle fibres.
iv. The chromosome appear as V,U or J shaped.

Telophase:
i. Reconstruction or reorganization phase.
ii. Karyokinesis (Nuclear division) is followed by cytokinesis which is the division of cytoplasm.
iii. At the equatorial plate the origin of phragmoplast occurs which gives rise to cell plate and subsequently to cell wall.
iv. Phragmoplast develops from small vesicle of golgi body and interzonal spindle fibres which are formed by microtubules.
v. In plant cell division of cytoplasm begins in the centre of the cell and gradually extends outside in a plane.
vi. In animal cell division, cytokinesis takes place by cell furrowing in which a furrow starts in the middle of the cell wall and gradually deepens or extends up to the middle of the cell thereby dividing the cell into two equal parts.

vii. At places where cell wall remains incomplete some tubular connections appear between daughter cells which is called plasmodesmata.

Factors affecting mitosis:
- The highly specialized cells (Nerve cells) divide slowly compared to lower specialized cells (embryonic cells)
- Temperature
- The wound and damaged cells show rapid multiplication.

Meiosis (2n to n):
- In most animals the meiosis occur just before fertilization and results in the formation of haploid gametes.
- Meiosis was first demonstrated by Farmes and Moore in 1905.
- It include two divisions,
  a. Reductional division – resulting in two haploid cells
  b. Equational division: - Dividing two haploid cells into two identical or similar haploid cells
- Meiosis include two nuclear divisions and one chromosome division.
- There are three types of meiosis based on variation in time and place of occurrence. They are,
  a. Gametic or terminal meiosis (Diplotonic pattern) – In this type, the meiosis occur immediately before gamete formation. Cells transform directly into sperm and egg cell without further cell division. This type occur in lower plants nad animal cells.
  b. Zygotic or initial meiosis (Haplotonic pattern) – Gametes fuse to form zygote (2n) which is the only diploid stage in the life cycle. The zygote enter the meiosis immediately and divide into four haploid cells.
      Example: Thallophyta (primitive meiosis)
  c. Sporic or intermediate meiosis (Diplohaplotonic pattern) – There is an alternation between haploid and diploid generations. Fertilization results in the diploid sporophyte generation. Meiosis occurs in the spore formed. The haploid spores germinate to form gametophyte. Gametophyte stage is

marked by the production of gametes. The gametes fuse to form sporophyte and the cycle continues.

Pre mitotic interphase:

It is the S phase and is marked by chromosome duplication excluding only 0.3% of DNA, the duplication of which occur during Zygote stage of Prophase I

❖ First nuclear division

Prophase I

i. It consume long duration of time like few hours to few days
ii. Prophase I is divided into five substages based on the pairing between homologous chromosomes, condensation of chromosomes, crossing over between them, terminalisation (movement of Chiasma to the outside or distal end)

Leptotene or leptonema:

i. It is the condensation stage.
ii. Chromosome may appear as single but it is actually double (due to replication).
iii. Nuclear volume is increased due to RNA synthesis.
iv. Chromosomes become visible as granules.
v. In the microsporocytes of *Lilium* the chromosomes are visible in the form of bunches on one side of nucleus and generally associated with nuclear membrane and the remaining part of nucleus is left vacant. This peculiar arrangement is called Bouquet's stage or synizesis or synizetic knot.
vi. Half of the chromosomes are paternal and half are maternal.
vii. Chromosome number is haploid (n)

Zygotene or Zygonema:

i. Chromosomes become shorter and thicker due to continuous dehydration of water.
ii. The lateral association of chromosomes called synapsis takes place at the point of centromere. During synapsis the chromosomes are seen in bivalent condition (2n).

iii. The synaptonemal complexes appear. Protein formed between homologous chromosomes (1000 A°) make the crossing over easier.
iv. Pairing begins at the point of centromeres and proceed towards the end (procentric pairing) or may begin at the end and proceed toward the centromere (proterminal pairing).
v. Replication of remaining 0.3% DNA which are Z form of DNA.

Pachytene:
i. Each bivalent divide longitudinally into two sister chromatids.
ii. Two homologous chromosomes coil around each other like rational coiling.

Diplotene:
i. Homologous chromosome begin to repel.
ii. Nucleolus, nuclear membrane disappear
iii. The chromosome number is 2n
iv. The homologous chromosomes do not separate completely and remain attached at points called chiasmata.
v. Segmental interchange of nonsister chromatids occur which is termed as crossing over. Crossing over was first discovered by Stern in *Drosophila*.
vi. Synaptonemal complex disappears.

Diakinesis:
i. The only difference from diplotene is that it have highly contracted bivalents and shaped in the form of X,O,V or loops.
ii. Chiasmata move from centre to peripheral ends termed as terminalisation.

Metaphase I
i. Chromosomes in equatorial plate and centromere of the homologous chromosomes point towards the opposite poles and chiasmata lie on the equatorial plane.
ii. Spindle fibres become visible and get attached to the centromere.

iii. The spindle fibres are of three types named chromosomal, continuous and interzonal fibre:
   a) Chromosomal – the fibres extend from the centromere to the pole of the spindle. The fibres become shortened and pull the daughter chromosomes apart.
   b) Continuous – the fibres extend from pole to pole without attaching the chromosomes. They elongate during cell division and push the spindle poles apart.
   c) Interzonal – The fibres are formed between poles regardless of whether the chromosome is present or not.

Anaphase I
i. Movement of one half of the chromosome to one pole and the other half to another pole.
ii. The two chromatids are united by a centromere.
iii. Movement of chromosomes to poles is random.
iv. In homologous chromosomes, the chromatid situated outward remain unchanged while the one inward contains a mixture of paternal and maternal segments.
v. Centromeres remain undivided and the chiasmata disappear.
vi. The chromosomes do not separate at a time, short chromosomes separate quickly whereas long chromosome separation is delayed due to the presence of interstitial chiasmata.

Telophase I
i. Arrival of chromosomes to the opposite poles.
ii. Chromosome begin to uncoil
iii. Nuclear membrane and nucleolus reappear.
iv. Cytokinesis followed by cell plate formation results in two daughter cells. In certain cases cytokinesis occur only at meiosis II

Meiosis I is directly followed by Meiosis II

Meiosis II *or second nuclear division:*

Prophase II
i. Unike Prophase I it occurs for a relatively short duration.

ii. Coiling, contraction of chromosomes and disappearance of nucleoli and nuclear membrane.

Metaphase II
i. Chromosomes get arranged at equatorial plate.
ii. Spindle fibre formation occurs and get attached to the centromere.

Anaphase II
i. Centromeres divide longitudinally and sister chromatids begin to separate and move from centre to poles

Telophase II
ii. Nucleolus and nuclear membrane reappear.
iii. Cytokinesis occurs resulting in the formation of four haploid daughter cells.

DIFFERNECES BETWEEN PLANT AND ANIMAL CELL MITOSIS

Animal Cell Mitosis	Plant cell Mitosis
Cell become rounded before division	No shape change
A number of hormones induce cell division and not a specific hormone	Cell division is induced by a specific plant hormone- cytokinin
Centrosome is essential for it	Centrosome do not occur
Mitotic apparatus contains asters*, spindle is amphiastral*	Mitotic apparatus is without aster and the spindle is anastral*.
Spindle degenerate at the time of cytokinesis	Most of the spindles persist as phragmoplast
Midbody may be formed during cytokinesis	Midpoint is absent
Cytokinesis occurs through cleavage	Cytokinesis occur commonly by cell plate method.
Microfilaments are involved in cytokinesis	Microfilaments have no major function in mitosis
Cleavage proceeds centripetally	Cell plate grows centrifugally

A furrow is formed between two daughter cells	Solid middle lamella develop in between two daughter cells for permanent adhesion
It occurs in bone marrow and many epithelia	It is found in meristem

*Aster: It is a star shaped cellular structure formed around centrosome during mitosis. It is composed of microtubules, radiare from centrospheres and look like a cloud (Amphiastral). Anastral implies the absence of asters in cells, this condition occur in cells without centrioles.

Leader sequence (mRNA)

- The 5' untranslated region (5'UTR) is referred to as leader sequence or leader RNA.
- It is the region of mRNA that is directly upstream from the initiation codon.
- This region is important for the regulation of translation of a transcript by different mechanisms in viruses, prokaryotes and eukaryotes.
- While called untranslated, the 5'UTR is sometimes translated into protein product.
- This product then regulates the translation of main coding sequence of mRNA.
- In many, 5'UTR is completely untranslated instead forming complex secondary structure to regulate translation.
- 5' UTR is related to interact with proteins relating to metabolism.

Splicing

In molecular biology, splicing is the editing of the nascent precursor mRNA transcript into mature mRNA. After splicing introns are removed and the exons are joined together.

Trans splicing

Trans splicing refers to the joining of two transcripts. Transplicing is a special form of RNA processing in eukaryotes where exons from two different primary RNA transcripts are joined end to end and ligated.

Alternate Splicing

It is a regulated process that results in single gene coding for multiple proteins. In this, particular exons of a gene may be included within or excluded from the final processed mRNA produced from that gene. The proteins translated from alternatively spliced m RNAs will contain differences in their amino acid sequence.

Rna editing

RNA editing is a molecular process through which some cells can make discrete changes to specific nucleotide sequences within a RNA molecule after it has been generated by RNA polymerase.

Special forms of RNA

Small nucleolar RNA (sno RNAs)

It is a class of small RNA molecules that primarily guide chemical modifications of other RNAs, mainly rRNA, tRNA and small nuclear RNAs.

There are two main classes of Sno RNA. They are C/D box snoRNA which enhance methylation by facilitating the attachment attachment or substitution of methyl group onto various substrates. The second class is H/ACA box snoRNAs responsible for pseudouridylation which is the conversion of nucleotide uridine to different isomeric form pseudouridine (Ψ).

Mature human rRNA contain approximately 95 Ψ modification.

Small cajal body RNA (Sca RNAs)

They resemble SnoRNAs and play similar role in RNA maturation but their targets are spliceosomal snRNAs and they perform site specific modifications of spliceosomal snRNAs precursors in the cajal bodies of the nucleus.

Small interfering RNA or silencing RNA (SiRNA)

These are double stranded RNA molecules, 20-25 base pairs in length and operating within the RNA interference pathway. It interferes with the expression of specific genes with complementary nucleotide sequences by degrading mRNA after transcription resulting in no translation.

Small nuclear ribonucleic acid (Sn RNA) or U-RNA

These are found within splicing speckles and cajal bodies of cell nucleus having a length of about 150 nucleotides. Their main function is the processing of pre-messenger RNA in the nucleus and also aids in the regulation of transcription factors. They play a sufficient role in maintaining telomeres.

TRANSLATION IN EUKARYOTES

Translation is the synthesis of protein from aminoacids coded from mRNA. Protein synthesis takes place in the ribosomes and hence the site for protein synthesis is cytoplasm where the ribosomes are located. Eukaryotic ribosomes are 80s with 40s and 60s subunits.

> 60+40 is not equal to 80. In the case of ribosomes we are not doing mathematical calculation. The numbers 60,40,80 are Svedberg units which is the non metric unit for sedimentation rate. Thus, the eukaryotic ribosomes have two subunits. The larger subunit sediments at 50s and the smaller subunit sediments at 40s, but the two together sediments at 80s. Same in the case of prokaryotes where the two sub units 50s and 30s together sediments at 70s

mRNA is synthesized from DNA by a process called transcription. In eukaryotes there is only one initiation and termination sites. mRNA sequence is used as a guide for the synthesis of protein. The main components required for translation are mRNA, rRNA, ribosomes, 20 kinds of aminoacids and their specific tRNAs. There are two types of factors involved in translation of eukaryotes:

Initiation factors : eIF2,eIF3,eIF4A, eIF4E, eIF4F and eIF4G

Elongation factors: EF1, EF2

There is a single release factor RF for recognition of three termination codons (UAA, UAG, UGA)

Enzymes involved

In eukaryotic translation two types of enzyme are involved. Aminoacyl tRNA synthetase, an enzyme that catalyse the bonding between specific tRNAs and the amino acid. The enzyme peptidyl transferase connect A site and P site by forming a peptide bond (the nitrogen carbon bond) during elongation phase.

Steps in protein synthesis- overview

Ribosomes are composed of 2 subunits the larger and the smaller subunits, 60s and 40 s respectively. There are four binding sites on ribosomes; one site for the binding of mRNA and the remaining three sites for binding tRNA. The three sites are P,A,E sites. The P site known as Peptidyl site binds to the tRNA holding the growing peptide chain of aminoacid. The A site is the acceptor site that binds to aminoacyl tRNA which holds to new aminoacid to be added to the polypeptide chain. The E site is the exit site where tRNA after releasing its aminoacid is let go by the ribosome.

Once the small subunit associates with mRNA molecules, the two subunits come together, creating a compactor that keeps tRNAand mRNA in stable and proper orientation for protein synthesis.

Every amino acid has terminal nitrogen group on one end and the carboxyl groups on the other end. The aminoacid is transferred from the aminoacyl tRNA in the A site to the growing protein chain attached to the Psite, they orient in specific direction such that chain grows by adding aminoacid to the carboxyl end of the chain. In this way protein chain grows from the nitrogen to carboxyl direction thus forming polypeptide chain. Each aminoacid is called a peptide.

There are two types of codons named start and stop codons. AUG (methionine) is the start codon as it initiates the process of translation and one of the three stop codons UAA(amber), UAG(ochre), UGA(opal) cause the termination of polypeptide chain thus terminating translation.

In eukaryotes there is no overlapping of transcription and translation.

Unwound DNA

In DNA, the sugar and bases are not planar. Long unwound DNA are found in nucleus during interphase. While unwinding DNA, the A + T rich sequence melt first. In the presence of super helical energy (a high energy state of DNA resulting from its supercoiling which is the natural form of DNA in the chromosome of most organisms), A+T rich regions can unwind and remain unwound under conditions normally found in the cell such sites often provide places for DNA replication proteins to enter DNA to begin the process of chromosome duplication

Cruciform structure:

DNA sequences are said to be Palindrome when they contain inverted repeated symmetry as in the sequence GGAATTAATTCC, reading from 5' to 3' end palindrome sequences can form intramolecular bonds (within single strand), rather than normal intermolecular (between two complementary strands), hydrogen bonds. To form cruciform (cross shaped) DNA must form small unwound structure, and the base pair must begin to form within each individual strand, thus forming four stranded cruciform structure.

Slipped Strand DNA

It can form within direct repeat DNA sequences such as (CTG)n.(CAG)n and(CGG)n. (CCG)n. Here n denotes variable number of repetitions.

Nuclear receptors

Nuclear receptors in molecular biology are modular in structure and contains the following domains:

A/B	C	D	E	F

A/B N- terminal regulatory domain

It contains the activation function 1(AF1) whose action is independent of the presence of ligand. The transcriptional activation of AF 1 is normally very weak, but it does synthesize with AF 2 in the E- domain to produce a more robust upregulation of gene expression. The A/B domain is highly variable in sequence between various nuclear receptors.

DNA *binding domain* (DBD)

It is a highly conserved domain (region c) containing two Zinc fingers that binds to specific sequences of DNA called Hormone Response Elements (HRE)

Hinge region

It is thought to be a flexible domain that connects the DBD with LBD. It influences intracellular trafficking and sub cellular distribution.

Ligand binding domain (LBD)

It is moderately observed in sequence (region E) and highly conserved in structure between the various nuclear receptors. The structure of the LBD is referred to as an alpha helical sandwich fold in which three anti parallel alpha helices (the sandwich filling) are flanked by two alpha helices on one side and three on the other(the bread). The ligand binding cavity is within the interior of LBD and just below three anti parallel alpha helical sandwich filling. Along with the DBD, the LBD contributes to the dimerization interface of the receptor and in addition, binds co activator and co repressor proteins. The LBD also contains the activation function 2 (AF2) whose action is dependent on the presence of bound ligand.

C- Terminal domain

It is highly variable in sequence between various nuclear receptors.

Molecular markers

The use of tools such as molecular markers or DNA fingerprinting can map thousands of genes. The screening is based on the presence or absence of a certain genes as determined by laboratory procedure rather than on the visual identification of the expressed trait in plants. Molecular markers can be used for preparation of linkage map, map based Colony, to map quantitative trait loci (QTL) for indirect selection of traits diagnosis of human diseases etc.

Tryptophan operon

In tryptophan operon the structural genes are regulated in two ways:

- By trp repressor- the presence of tryptophan causes the repression of trp operon
- Attenuation – the presence of tryptophan cause the formation of attenuator terminator which decoupies transcription and translation

Puromycin

Puromycin is an antibiotic used to inhibit protein synthesis. It resembles aminoacyl tRNA and it can bind to the ribosomal A site and participate in peptide bond formation. The product of this reaction, instead of being translocated to the Psite, dissociates from the ribosome, causing premature chain termination. It inhibits both prokaryotic and eukaryotic protein synthesis.

Statements on Eukaryotes DNA replication:

Eukaryotes often have multiple origin of replication on each linear chromosome that initiate at different times (replication timing) with upto 100000 present in a single human cell. Having many origins of replication helps to speed the duplication of their much larger store of genetic material. The segment of DNA that is copied starting from each unique replication origin is called a replicon. During the embryonic S-M cycles, entry into S phase must be controlled post transcriptionally and must be regulated in the presence of constitutive cyclin E kinase activity. The addition of G1 phase during embryogenesis requires that extrinsic development cues influence the onset of S phase. In *Drosophila*, this is mediated atleast in part through cyclin E. During the S-G cycle that produces polytene cells, entry into S phase can no longer have the completion of mitosis as a prerequisitic. In addition to differential control of the onset of S phase, the actual parameters of S phase are altered during development. By parameters of S phase we mean the intrinsic properties of DNA replication. The parameters of DNA replication changed in modified cell cycles include replication origin usage and activation, the rate of replication fork movement and the block to rereplication. By pulse labelling cells with nucleotides prior to this treatment, it is possible to directly examine sites of DNA synthesis. Originally used with radioactively labelled nucleotides to accurately measure the rate of replication fork progression, more recent studies have used this approach to compare the efficiency of origin firing between early and late replicating regions.

The dihedral angles of 20 residue peptide are represented in the Ramachandran plot. Can we conclude that the peptide doesn't have a proline?

Proline is mainly present in β-structure but is also predominant in the n-terminal of α helix. So we can't conclude using Ramachandran plot particularly whether proline is present or absent in the secondary structure of protein.

The second messenger cAMP synthesized by adenylyl cyclase transduces a wide variety of physiological signals in various cell types in mammalian cells. Most of the diverse effects of cAMP are mediated through activation of protein kinase A (PKA) also called as cAMP dependent protein kinase.

This is because in active form PKA have four units (tetramer)of which two are receptor unites and the remaining two are catalytic units. Each receptor has 2 sites where cyclic AMP can bind. In the presence of cAMP, receptor site release the catalytic sites and convert it into active form.

While studying the binding of proteins to the cytoplasmic face of cultured liver cells one have found a method that gives a good yield of inside out vesicles from the plasma membrane. Unfortunately the preparations are contaminated with variable amounts of rightside out vesicles. Somebody suggests that passing the vesicles over an affinity column made of lecithin coupled to sepharose beads will avoid this contamination. How?

Lecithins found in all organisms are proteins that bind carbohydrate with high affinity and specificity. Glycoproteins are present on the extracellular surface of the plasma membrane which form the right side out vesicle. So lecithin will specifically bound to these glycoproteins but the cytoplasmic phase of the plasma membrane do not have glycoproteins due to which lecithin will not bound to inside out vesicles.

During receptor-mediated endocytosis, apolipoprotein B on the surface of LDL particle binds to the LDL receptor present in coated pit containing clathrin. The receptor LDL complex is internalized by endocytosis, trafficked by lysosome and the LDL receptor is finally recycled. A patient reports with familial hypercholesterolemia. This is due to:

Cholesterol and cholesteryl esters moved from the tissue of origin to the tissue in which they will be stored or consumed in the blood plasma as plasma lipoprotein. Macromolecular complexes of specific carrier proteins called apolipoproteins with various combinations of phospholipids, cholesterol etc. The cholesterol enters cells by receptor mediated endocytosis called LDL receptors. A mutation in LDL receptor results in a condition termed Hypercholestrolemia.

Budding yeast cells that are deficient for MAD 2, a component of spindle attachment check point are killed by treatment with benomyl, which causes microtubules to depolymerise. In the absence of benomyl, however the cells are perfectly viable. Reason?

MAD 2 (Mitotic arrest deficient 2) is an essential spindle checkpoint protein. The spindle checkpoint system is a regulatory system that restrains progression from the metaphase to anaphase transition. The MAD 2 gene was first identified in the yeast *S. Cerevisiae* in a screen for genes which when mutated would confer sensitivity to microtubule poisons. Wild type cells arrest in mitosis in response to spindle damage when treated with nocodazole or benomyl but in its absence, spindles are formed normally.

Eukaryotic genomes are organized into chromosomes and can be visualized at mitosis by staining with specific dyes. Heat denaturation followed by staining with Giemsa produced alternate dark and light bands. Explain?

The staining of mitotic chromosomes distinguished the regions of DNA that is rich in A-T nucleotide pairs (G bands) from the regions of DNA that is rich in G-C nucleotide pair (R bands). The G bands strain dark with Giemsa strain while R bands are not. Both G and R bands are known to contain genes. The dark bands obtained by this process are mainly AT rich and gene rich regions.

Lac repressor inhibits expression of genes in Lac operon whereas Purine biosynthesis is repressed by the Pur repressor. The two proteins have 31.2% identical sequences and have similar three dimensional structures. The regulatory properties of these proteins differ in relation to:

- *Binding of small molecules to the repressor*
- *Presence of recognition sites on the genome*

This is because the repressor is a proteinaceous substance synthesized by the regulator gene which blocks the operator gene so that the structural genes are unable to form mRNA by transcription and thus protein formation is halted. The presence of 31% identical sequence in lac repressor and pur repressor do not confirm same regulatory properties because of binding of small molecule to repressor (inducer) and the recognition site in the genome.

Two E.coli cultures A and B are taken. Culture A was earlier grown in the presence of optimum concentration of gratuitous inducer IPTG. Both the cultures are now used to inoculate fresh medium containing sub optimal concentration of gratuitous inducer. It was observed that culture B was unable to synthesize lactose, whereas culture A did so efficiently.

Reason: When the optimum concentration of gratuitous inducer IPTG is present in a medium in which bacterial cells are growing, this interact with the repressor and make it inactive. The inactive repressor is unable to bind to operator and hence the transcription for lactose permease takes place. This allows preferential uptake of lactose. In short, in culture A, lactose permease was induced to a high level, during pretreatment with IPTG, which allowed preferential uptake of lactose.

It has been observed that in 5-10% of the eukaryote mRNAs with multiple AUGs, the first AUG is not the initiation site. In such cases, the ribose skips over one or more AUGs before encountering the favourable one and initiating translation. This is postulated to be due to the presence of sequences CCACCAUGG and CCGCCAUGG.

Reason: In eukaryotes, the ribosomal small subunit first binds at 5' end of mRNA chain and added by recognition of 5'cap. This subunit then propels itself along the mRNA chain in scanning mode, in search of an AUG codon. If this recognition site is poor enough scanning ribosomal subunit will ignore the first AUG codon in the mRNA and skip to the second or third AUG codon instead. This phenomenon is known as Leaky scanning.

Presence of circular mRNA for a specific protein in an eukaryotic cells reflects a rapid rate of synthesis of that protein.

Reason: The Poly (A) binding protein PAB iP interacts directly with eukaryotic translation initiation factor 4G (eIF4G) to facilitate translation initiation of polyadenylated mRNAs in yeast. Although the eIF4G-PABp interaction has also been demonstrated in mammalian system but its biological significance in vertebrates is unknown. There are several possible roles of PABp/eIF4G interaction to stimulate translation (1) promoting ribosome recycling (2) Stimulationg 60s ribosome joining (3) increasing the affinity of eIF4F for the cap (viz., 40s ribosome recruitment).

Si RNA and miRNA are used for achieving gene silencing. Although major steps are similar there are distinct differences in the key players of the two processing pathways:

Small interfering RNA (siRNA) also called short interfering RNA or silencing RNA, is a class of double stranded RNA molecules, 20-25 nucleotides in length. siRNA plays many roles, but its most notable role is in the RNA interference (RNAi) pathway where it interferes with the expression of specific genes with complementary nucleotide sequence. Argonaute proteins are the catalytic components of the RNA-induced silencing complex (RISC), the protein complex responsible for gene silencing phenomenon known as RNA interference (RNAi). Argonaute proteins bind different classes of small non-coding RNAs including micro RNAs (miRNAs), Small interfering RNA (siRNAs) and Piwi interacting RNA (piRNAs). Small RNAs guide argonaute proteins to their specific targets through sequence complementarity, which typically leads to silencing of the target. Some of the argonaute proteins have endonuclease activity directed against mRNA strands that display extensive complementarity to their bound small RNA and this is known as slicer activity.

Glucose is mobilized in muscle when epinephrine activates Gas. In an experiment in which muscle cells were stimulated with epinephrine, glucose mobilization was observed even after withdrawal of epinephrine. This could be due to the presence of cAMP phosphodiesterase inhibitor.

Reason: The hormone epinephrine is important in increasing plasma glucose concentration during stress. Many hormones act through a signalling cascade, a series of steps, in each of which a catalyst activates a catalyst resulting in very large amplification of original signal. Epinephrine activates adenylyl cyclise which produces many molecules of cAMP for each molecule of receptor bound hormone. cAMP inturn activates cAMP dependent protein kinase which activates phosphorylase kinase that activates glycogen phosphorylase. The result is signal amplification. One epinephrine molecule causes the production of many thousands of molecules of glucose-1 phosphate from glycogen. In tissues with α_2 adrenergic receptors epinephrine lowers the cAMP because the α_2 receptors are coupled to adenylyl cyclase through an inhibitory G-protein (Gi)

In eukaryotic chromatin, 30 nm fibre (solenoid) can open up to give rise to two kinds of chromatin. In one type (A), the promoter of a gene is within the open chromatin is occupied by a nucleosome whereas in the other (B), the promoter is occupied by histone H1. From this it is understood that the gene (A) is repressed and the gene (B) is active.

Reason: A nucleosome is composed of 146 nucleotide pairs long bound to histone octamer of H_2A, H_2B, H_3 and H_4 (2 each). Each nucleosome is separated from the next by a linker DNA when the promoter of a gene within the open chromatin is occupied by nucleosome it will be repressed but when occupied by histone H1 It will be active.

> A nucleosome is a basic unit of DNA packaging in Eukaryotes, consisting of a segment of DNA wound in sequence around eight histone protein cores. This structure is often compared to thread wrapped around a pool.

Genetic studies demonstrated that TBP *mutant cell extracts are deficient in transcription of genes from all three promoters viz. Class* I,II,III. *The characteristics features of* TBP *are:*

The TBP gene provides instructions for making a protein called TATA box binding protein. This protein is active in cells and tissues throughout the body, where it plays an essential role in regulating the activity of most genes. The TATA box binding protein attaches to a particular sequence of DNA known as the TATA box. This sequence occurs in a regulatory region of DNA near the beginning of many genes. Once the protein is attached to the TATA box near a gene, it acts as a landmark to indicate where other enzymes should start reading the gene. The process of reading a gene's DNA and transferring the information to a similar molecule called mRNA is known as transcription. One region of the TBP gene contain a particular DNA segment known as CAG/CAA trinucleotide repeat. This segment is made up of a series of three DNA building blocks (nucleotide) that appear multiple times in a row normally the CAG/CAA segment is repeated 25-42 times within the gene. The assembly process in a transcription start with the binding of TFIID to the TATA sequence, a short double helical DNA sequence primarily composed of T and A nucleotides. TFIID is composed of many subunits, that are responsible for recognizing the TATA sequence called TBP. TBP participate in the initiation of transcription of all genes. Once TFIID is bound to this DNA site, the other general transcription factors along with RNA Pol II are added in turn.

Proteins in the cells can be visualized by expressing the gene (coding for the said protein) as a fusion with the green fluorescence protein (GFP) *and directly visualize under a fluorescence microscope.*

FRET microscopy is a powerful and popular approach to study protein interactions in living cells. Green fluorescent protein (GFP) is isolated (like aequorin) from the jellyfish *Aequoria Victoria*. The freshly translated protein is not fluorescent but within an hour or so it undergoes self catalyzed post translational modification to generate an efficient and bright fluorescent centre, shielded within the interior of a barrel like protein.

The following are the methods employed in detecting protein:

- Microscopy, protein immunostaining
- Protein immunoprecipitation
- Immunoelectrophoresis
- Immunoblotting
- BCA protein assay (to measure protein concentration)
- Western blot
- Spectophotometry
- Enzyme assay.

For the generation of transgenic plants in crop improvement, one important regularity gene X was over expressed in a crop plant. Out of

30 transgenic rice plants generated 22 showed high levels of gene X expression. However rest 8 lines displayed low levels of expression. Such observation is Gene silencing effect

Theory: It is now possible to perform site directed changes in genes and to synthesize entirely new genes. Transgenic organisms can also be produced to inactivate a particular gene in which case it is known as knockout technology. The gene expression falls in gene silencing. Certain genes which are turned on and off depending on the cellular requirements are sometimes kept inactive by a mechanism called repression.

There are many genes in the organisms which are expressed at varying levels under different conditions or in different tissue. This is called suppression effect. The phenomenon referred to as "gene co-suppression" has been studied most extensively in dicot plants. In some cases of co suppression, methylation (and thus suppression of transcription) of the endogenous gene has been implicated. In other cases, methylation is not detected. The gene is being actively transcribed, but no RNA accumulates. The genes (and viruses) are said to be victims of the post transcriptional gene silencing

If one wishes to design a microarray chip for whole genome expression analysis of an eukaryotic system, the region preferred for the selection of unique target sequence is 3'region of the CDs and 3'untranslated region (UTR)

Reason: DNA microarrays have DNA molecules immobilized at precise locations on glass or silicon surface. To design a microarray chip for whole genome expression analysis 5' region of coding DNA sequence and 5' end of untranslated region should be preferred for selection of unique target sequence.

For 5'end labelling of DNA,*the following reactions are carried out sequentially as indicated 5' dephosphorylated* DNA+[γ^{32}P]dATP+T$_4$ *polynucleotide kinase* (T$_4$PNK) *and incubated for 2 hours in ammonium*

acetate followed by Tris EDTA and ethanol. If trace amount of NH_4^+ is present in the initial DNA mix, it inhibits T_4PNK, therefore should not be present in the DNA mix.

Reason: This procedure is useful for radioactive labelling of oligonucleotides. The 5' end of DNA should be dephosphorylated with alkaline phosphatase before kinase treatment. The alkaline phosphatise treatment can also be used to prevent recircularization and relegation of linearized. Ammonium ions are strong inhibitors of bacteriophage T_4 polynucleotide kinase; therefore DNA should not be dissolved in or precipitated from, buffers containing ammonium salts prior to treatment with kinase.

Three E.coli mutants are isolated which require compound A for hair growth the compounds 'A' for their growth. The compounds B,C,D are known to be involved in the biosynthetic pathway to A. In order to determine the pathway, the mutants were grown in a medium supplemented with one of the compounds, A to D the results are summarised below,

Mutant	Medium supplemented with compound			
	A	B	C	D
1	+	0	0	0
2	+	0	0	+
3	+	0	+	+

+ indicates growth and 0 indicates lack of growth. The equation that represent the biosynthetic pathway of A is

B $\longrightarrow$ C $\longrightarrow$ D $\longrightarrow$ A.

Reason: All the three mutants do not show growth in presence of B but strain 3 undergoes growth in the presence of C, strain 2 and 3 in the presence of D and all the 3 mutants in the presence of A showed that the biosynthetic pathway involved the sequence

B ⟶ C ⟶ D ⟶ A.

Upon ligand binding, cell surface receptors laterally to be capped and internalized Leishmania, a protozoan parasite, can use several receptors on macrophages to get internalized. One of them is Toll like receptor 2 (TLR2) that binds lipophosphoglycan on Leishmania. Once internalized, the parasite is destroyed in the phagolysosome. Treatment of Leishmania infected macrophages with β- MCD and ammonium chloride will result in lowest parasite number in macrophages.

Theory: *Leishmania donovani* is an obligate intracellular parasite that infects macrophages of the vertebrate host, resulting in visceral leishmanias in humans, which is usually fatal if untreated. Cholesterol depletion from macrophage plasma membrane using methyl β cyclodextrin (MβCD) results in a significant reduction in the extent of Leishmanial infection. Furthermore, the reduction in the ability of the parasite to infect host macrophages can be reversed upon replenishment of cell membrane cholesterol cyclodextrin are the oligomer with seven residues of methylated glucose, has been extensively used to selectively and efficiently extract cholesterol.

The Lac Operon in E coli is controlled by both the lac repressor and the catabolic activation protein CAP-*In an invivo experiment with Lac Operon the following observations are made:*

- cAMP *levels are high*
- *repressor is bound with allolactose*
- CAP *is interacting with* RNA *polymerase*

The most appropriate conclusion made is that glucose is absent and lactose is present.

Reason: The regulatory mechanism of Operon is responsible for utilisation of lactose as a carbon source is called Lac Operon. In Lac Operon three structural genes (Z, Y, A) are associated with lactose utilization. However the Lac Operon cannot function in presence of sugar other than lactose allolactose binds with lac repressor protein to form inducer repressor complex. Consequently repressor is released from lac O due to changes in 3D confirmation. After being free lac O allows the RNA polymerase to form mRNA. When E.coli grown in medium containing glucose the cAMP concentration in cell falls down but in alternate carbon source cAMP level is increased, which inturn activates CAP proteins.

During cell cycle regulation in eukaryotes there are post translational modifications of proteins factors which act as switches for different phases of cell cycle. A cell population of yeast was transfected with gene for weekinase (modifies cdk_2 *protein).*

The cell cycle control system is based on two key families of protein the cdk which induce downstream process by phosphorylating selected proteins on serine, thereonine and cycline that bind to cdk molecules and control their ability to phosphorylate appropriate target proteins. As the transfection efficiency in yeast cell production is only 50% then after division more than 50% cells have mutant modified cdc_2 protein. cdk_1 and cdk_2 are phosphorylated by weelkinase at the inhibitory residue near the N-terminus so the cells are arrested at G_1 phase. Dephosphorylation of these deciduous is performed by cdc 25 family of phosphatases.

A synthetically prepared mRNA *contains repetitive sequences. The* mRNA *was incorporated with mammalian cell extract which contains ribosomes,* tRNA *and all the factors required for protein synthesis. Assuming no initiation codon is required for protein synthesis, the peptide most likely to be synthesized is a single peptide composed of the same amino acid sequence.*

Reason: The repetitive AU sequence in mRNA coded for only two types of amino acids.

AUAUAUAUU

But in absence of initiation codon individual amino acids are coded for each peptide.

In semi conservative model of DNA *replication two parental strands unwind and are used for synthesis of new strands following the rule of complementary base pairing. Synthesis of complementary strands requires that DNA synthesis proceeds in opposite direction while the double helix is progressively unwinding and replicating in only one direction. One of the DNA strands is continuously synthesized in the same direction as the advancing replication fork and is called leading strand whereas the other strand is synthesized discontinuously in segments and is referred to as lagging strand. These short fragments made discontinuously are labelled as okazaki fragments. These okazaki fragments need to be matured into continuous* DNA *strand by the enzymes* DNA *polymerase 1 and* DNA *ligase.*

Reason: DNA polymerase carry out the process of polymerization of nucleotides and it can elongate the okazaki fragments. DNA ligase seals single strand nick in DNA. It catalyse the formation of phosphodiester bonds between 3'OH and 5'POH group of the nick.

Double stranded DNA *replicates in a semi conservative manner. In an in vivo* DNA *synthesis reaction, dideoxy* CTP *and dideoxy* CMP *were individually added in excess in separate reaction tubes in addition to* dNTPs *and other necessary reagents. Rate of* DNA *synthesis was measured by incorporation of* ^{3}H- *thymidine. The result is as follows:*

Each chain of the double helix acts as template and is involved in replication of DNA. Each purines and pyrimidines base of the strand forms hydrogen bonds with complementary free nucleotides to be involved in polymerization in cell. The rate of DNA synthesis is directly influenced by dideoxy CTP and dideoxy CMP.

The respiratory chain is relatively in accessible to experimental manipulation in intact mitochondria. Upon disrupting mitochondria with ultrasound, however it is possible to isolate functional sub mitochondrial particles which consist of broken cristae that have resealed inside out into small closed vesicles. In this vesicle the components that originally faced the matrix are now exposed to surrounding medium this arrangement helps in studying of electron transport and ATP synthesis because it is difficult to manipulate the concentrations of small molecules (NADH, ATP, ADP, Pi) *in the matrix of intact mitochondria.*

Reason: When mitochondria are examined in an electron microscope, the outside surface is seen to be studded with tiny spheres attached to the membrane by stalks. In intact mitochondria these lollipop like structures are located on the inner (Matrix) side of the inner membrane. Thus the sub mitochondrial particles are inside out vesicles of inner membrane with what was previously their Matrix facing surface exposed to the surrounding medium. As a result they can readily be provided with the membrane impermeable metabolites that would normally be present in the matrix space. When NADH, ADP and inorganic phosphate are added, touch preparations transport electrons from NADH to oxygen and couple this oxidation to ATP synthesis catalyzing the reaction ADP + Pi ⟶ ATP. This cell free system provides an assay that makes it possible to purify the many proteins responsible for oxidative phosphorylation in a functional form.

Cre/loxP *system is used by phage* P1 *to remove terminally redundant sequences that arise during packaging of the phage of the phage* DNA. Cre-lox *system can be used to create targeted deletions, insertions and inversion in genome of transgenic animals and plants. Consider a series of genetic markers* A *to* K. *How should the lox* P *sites be positioned* inorder that Cre recombinase can create an inversion in the EFG segment relative to ABCD and HIJK?

Answer: ABCD ⟶ EFG ⟶ HIJK

Reason: Inversion occurs when parts of chromosomes become detached turn through 180° and are reinserted in such a way that the genes are in reverse order. For inversion the lox P should be targeted within specific fragment of genes (EFG).

In 'Taq Man' assay for detection of base substitution (DNA variant) probe (oligonucleotides) with fluorescent dyes at the 5' end and a quencher at 3' end are used. While the probe is intact impact the proximity of the quencher reduces the fluorescence emitted by reporter dye. If the target sequences (wild-type or the variant) are present, the probe anneals to the target sequence, downstream to one of the primers used for amplifying the DNA sequence flanking the position of the variants. For an assay two flanking PCR Primers, two probes corresponding to the wild type and variant allele and labelled with two different reporter dyes and quencher were used. During extension the probe maybe cleaved by the taq polymerase separating the reporter dye and the quencher. 3 individuals were genotyped using this assay. Sample for individual I shows maximum fluorescence for the dye attached to the wild type probe, same for individual II shows maximum fluorescence for the dye attached to the variant probe and sample for individuals III exhibits equal fluorescence for both the dyes, so individual 2 is homozygous for variant allele.

Reason: Since the sample for individual II shows maximum fluorescence for dye attached to variant probe it is clear that variant allele (due to base substitution) is found in higher concentration or amount and it is homozygous.

In an invitro experiment using radiolabelled nucleotides, a researcher is trying to analyse the possible products of DNA replication by resolving the products using urea polyacrylamide gel electrophoresis. In one experimental setup RNase H was added (set 1), while in another set no RNase H was added (set 2). The possible observations are, there is a distinct difference in the mobility of the newly synthesized labelled DNA fragments between set 1 and set 2 and the mobility of the newly synthesized labelled DNA fragments in case of set 1 is faster as compared to set 2.

Reason: RNAase H is a non specific endonuclease which cleaves RNA via hydrolytic mechanism and responsible for removing RNA primer during DNA replication for completion of newly synthesized DNA strand. Presence of RNase H in set I degrade RNA and replaced with DNA which decreases the weight of newly synthesized DNA fragments.

Synthesis of normal haemoglobin requires coordinated synthesis of α globin and β globin. Thalassemia are genetic defects perturbed in this coordinated synthesis. Patients suffering from deficiency of βglobin chains (β-thalassemia) could also be due to mutations affecting the biosynthesis of βglobin mRNA. Mutations in the promoter region of the β globin gene. Mutation in the splice junction of the β globin gene. Mutations towards the 3'end of the β globin gene that codes for polyadenylation site. All these three statements describe the genesis of non functional β globin leading to β- thalassemia.

Reason: Introns are non coding regions of gene. An example of transcriptional control mutation in the β globin gene occurs at position 28. In his TATA box sequence has been changed from ATAAAA to ATACAA resulting in promoter down phenotype. The patient has severe beta plus thalassemia. In most mammalian genes the sequence AATAAA occurs near the 3'end of gene where it acts as a signal to terminate and add poly A. A patient has the sequence AATAAG in the two of his α globin alleles.

Pre mRNAs are rapidly bound by snRNAs *which carry out dual steps of* RNA *splicing that removes the intron and joins the upstream and downstream exons. Almost all introns begin with* GU *and end with* AG *sequences and hence all the* GU *or* AG *sequences are spliced out of* RNA. U$_2$ RNA *recognizes important sequences at the 3'acceptor end of the intron.*

Reason: The coding RNA sequence on either side of an intron sequence are joined to each other after the intron sequence has been cut out by a process called RNA splicing. Introns range in size from 80 – 10000 nucleotides most of the introns begin with GU and ends with AG RNA splicing is catalyzed by spliceosome formed from the assembly of U1,U2,U5,U4/U6 in snRNAs. The introns are excised in the form of a lariat according to splicing pathway.

For continuation of protein synthesis in bacteria, ribosomes need to be released from the mRNA *as well as to dissociate into subunits. These processes do not occur spontaneously. They need the following conditions;* RRF *and* EF-G *which aid in the process and* IF3 *and* IF1 *promote dissociation of ribosomes.*

Theory: Initiation, elongation, termination, release factors are required for the protein synthesis in both prokaryotes and eukaryotes

IF1 – Assist IF3 binding

IF-2 –Binds initiator tRNA and GTP

IF3- Release mRNA and tRNA from recycled 30s subunit and aid new mRNA binding

EFG – Promoter translocation through GTP binding and hydrolysis

RRF- Together with EFG induces ribosomal dissociation to small and large subunit.

Insulin and other growth factors stimulate a pathway involving a protein kinase mTOR *which in turn augments protein synthesis.* mTOR *essentially modifies protein (s) which in their unmodified form act as inhibitors of protein synthesis. The possible candidates are* eIF-4E-BP$_1$

Theory: The mTOR (mammalian target of Rapamycin) pathway involves the regulation of a protein kinase variously known as FKBP$_{12}$-rapamycin associated protein (FRAP) or rapamycin and FKBP$_{12}$ target 1(RAFT1). Activated mTOR phosphorylates protein phosphatise 2A(PP2A), S$_6$K$_1$ and eIF$_4$E-BP(PHAS$_1$). This phosphorylation of PP$_2$A prevents disphosphorylation of eIF$_4$E-BP. Phosphorylation of eIF$_4$E which then participate in formation of protein translation complex involving capped mRNAs.

Bacteriophage λ *has two modes in its life cycle, lytic and lysogenic. In the lysogenic mode, the expression of all the phage genes is repressed while the expression of repressor gene switches between on and off position depending on the concentration of repressor.*

Reason: The activity of C II protein bacteriophage lambda is probably the critical controlling factor in the choice of lytic or lysogenic pathway. The C I gene with the aid of C II gene product is transcribed from a promoter known as PRE (promoter repression establishment). Once CI is transcribed it is translated into λ repressor protein, it interacts at the left and right operators (O$_L$ and O$_R$). When these operators are bound by CI protein transcription of the left and right operon ceases. Removal of O$_L$ and O$_R$ or mutation in these leads to lytic cycle (virulent phage).

A *bacterial response regulators turn on gene A in its phosphorylated form. The amount of 'A' shows a sharp and steep rise at a threshold concentration of the signal sensed by the cognate sensor. This is likely to do increased phosphatase activity of the sensor at the threshold concentration.*

Reason: Although unicellular, bacteria are highly interactive and employ a range of cell to cell communication or Quorum sensing (QS) systems for promoting collective behaviour within a population. QS is generally considered to facilitate gene expression only when the population has reached a sufficient cell density and depends on the synthesis of small molecules that diffuse in and out of bacterial cells. As the bacterial population density increases, so does the synthesis of QS signal molecules and consequently their concentration in the external environment increases. Once a critical threshold concentration is reached a target sensor kinase or response regulator is activated, so facilitating the expression of QS dependent target genes. Several chemically distinct families of QS signal molecules have been described, of which the N-acythomoserine Lactone (AHC) family in gram negative bacteria have been the most intensively investigated. QS Contribute to environmental adaptation by facilitating the elaboration of virulence determinants in pathogenic species and plant bbiocontrol characteristics in beneficial species as well as directing biofilm formation and colony escape. QS also crosses the prokaryotic eukaryotic boundary in that QS signal molecules may directly facilitate bacterial survival by promoting an advantageous lifestyle within a given environmental niche.

Intracellular transport and cytoskeletal organization of a cell is regulated by nucleotide exchange of different small molecular weight Gtpase of Ras super family. Over expression of Rho in GTP bound form modulates the actin cytoskeleton of HeLa' cells.

Reason: The Rho family of GTPase is a family of small signalling G-protein and is a sub family of the Ras super family. The members of Rho GTPase family regulate many aspects of intracellular actin dynamics. The Rho proteins have been described a molecular switches and play a role in cell proliferation, apoptosis, gene expression and other cellular functions: microinjection of Rho protein into cultured cells leads to the appearance of large bundles of actin filaments known as stress fibres and to the enhancement of focal contacts, whose the cell is attached to the substratum externally and stress fibres are anchored internally.

Polynucleotide kinase (PNK) *is frequently used for radioactivity for radiolabelling* DNA *or* RNA *by phosphorylating 5'end of non phosphorylated polynucleotide chains.*

The reason is PNK is a T_4 bacteriophage encoded enzyme. PNK is inhibited by small amount of ammonium ions and has 3' phospahatase activity. PNK catalyzes the transfer of γ phosphate group added to the DNA at 5'end.

DNA *is not hydrolyzed by alkali whereas* RNA *is readily hydrolyzed due to the presence of* 2'OH *group in* RNA

Reason: RNA hydrolysis is a reaction in which a phosphodiester bond in the sugar phosphate backbone of RNA is broken, cleaving the RNA molecule, RNA is susceptible to this base catalyzed hydrolysis because the ribose sugar in RNA has a hydroxyl group at the 2' position. This feature makes RNA chemically unstable compared to DNA, which does not have their 2'OH group and thus is not susceptible to base catalyzed hydrolysis.

A *null mutation is created in a gene which is responsible for specific phosphorylation at 6th carbon position of mannose on acid hydrolases occurring in Cis Golgi. The effect of mutation if the acid hydrolases in the mutant do not get degraded is because the lysosomes will be devoid of lysosomal enzyme as it is secreted out.*

Null mutation: Null allele is a non functional copy of a gene caused by genetic mutation. Null mutation is a gene that usually encodes a specific enzyme leads to the production of non functional enzymes or no enzymes at all. Less severe mutations are called leaky mutations (leaks through). In the cis-golgi a GlcNAc phosphotransferase adds a GlcNAc-1-phosphate residue onto the 6-hydroxyl group of a specific mannose residue within thioligosaccharide. This forms GlcNAc. Once the phosphodiester has been formed the lysosomal enzymes will be translocated through the golgi apparatus to the trans golgi.

Inorder to prove that liposome can serve as a model membrane (mimicking cellular plasma membrane) and can be used as a target for complement mediated immunolysis an experiment is designed as follows. To initiate such experiment, hapten conjugated liposomes are made and loaded with umbelliferyl phosphate (UMP; *hydrolysed product of UMP is umbelliferone and is fluorescent). Such loaded, hapten conjugated liposomes in 10mM Tris buffered saline, pH 7.4 were mixed with anti hapten antibodies and fresh guinea pig serum (as a source of complement) to induce immunolysis of liposomal membrane. To quantify only the membrane lysis complement mixture is directly subjected to fluorescence measurement.*

Reason: The membrane is already lysed hence the UMP fluorescence will directly be measured in supernatant.

A gene producing red pigment was placed near centromeres of fission yeast and thus subjected to position effect variegation and produced white colonies. A screen for mutants that increased the red pigment production was undertaken. Histone deacetylase when mutated is likely to produce this genotype.

Reason: By the mutation of deacetylase gene caused to increased production of product. Normally deacetylase represses the gene expression but mutant deacetylase can't repress the gene expression.

If a proteasome inhibitor is added to synchronously cycling human cells in G_2 phase arrest cells in Anaphase occur.

Reason: The addition of proteasomes inhibitor to G_2 cells causes arrest of cells at anaphase. For metaphase to anaphase progression requires the activation of anaphase promoting factor (APF), which cause ubiquitination of two target proteins, one is securing and the other is cyclin B. Both securin and cyclin B are degraded by proteasome, hence cell enters into anaphase. So, inhibiting proteasome caused the cell arrest at anaphase.

Lysogenic cycle is more beneficial to a Virus than Lytic cycle because Lysogeny cause more mutations to occur in the virus, creating more variants upon which natural selection can operate

Theory: Upon penetration into an *E.coli* cell, the linear phage DNA forms a circle. The circle can recombine with and become part of circular bacterial DNA. The inserted phage DNA is now called a prophage. Every time the host cell machinery replicates the bacterial chromosome add prophage DNA. The prophage remains latent within the progeny cells. Under favourable conditions prophage is excised and it takes some portion of the host DNA. This event causes more mutation to occur in the Virus crearing more Variants.

The rate of mutation in E.coli from lac⁻ to lac⁺ is determined using a medium containing lactose, as the only source of energy. The principle of spontaneity of mutations can be said to be violated if in the presence of lactose, the rate of mutation from lac⁻ to lac⁺ is increased but overall rate of mutation is not.

Reason: An inducible enzyme β lactase is found in *E.coli* and hydrolyses lactose to glucose and galactose. When lac^+ *E.coli* cells are grown on nutrient medium devoid of lactose β galactosidase cannot be detected within the cell and such cultures are phenotypically lactose sensitive (Lac-). On the other hand if lactose is substituted for glucose *E.coli* cells quickly produce large quantity of enzymes and permease transportation lactose into cells when lactose is removed from the environment the cells stop producing β galactosidase and revert to phenotypically lac- population

Humans and chimps differ more in DNA *sequences of pseudo genes than in coding regions of functional genesis a prediction of the neutral theory of molecular evolution.*

Theory: Pseudo genes are often very similar to functional genes but they cannot be translated into proteins. The chemical composition of DNA is basically the same in living beings except for difference in sequences of nitrogenous bases. The degree of similarities between DNA of two species can be estimated by pairing property of DNA strands. Human and Chimpanzee have similarities in DNA but differ in coding sequences.

In a Radio immune Assay (RIA) for glucocorticoid hormone, radioactive glucocorticoid (tritiated) is added to the RIA cocktail. When the amount of bound hormone was measured no counts were observed because, the radioactive tag to the hormone completely dissociated during storage. Antibody was not added to the cocktail.

Theory: Radio immune assay (RIA) is a highly sensitive technique and can measure even the less concentration (0.001μg/ml) of antigen or antibody. Liquid phase RIA is based on competitive binding of radiolabelled and unlabelled antigen to a high affinity antibody. The increasing amount of antigen (unlabelled) of known concentration competes with radiolabelled for available site of antibody.. The amount of labelled antigen in solution is measured and the concentration of unlabelled antigen can be determined. If the radioactive antigen is sufficient or specific activity of tritium is low no counts were observed.

CONCEPTS RELATED TO BIOCHEMISTRY

Biosynthetic pathways of secondary metabolites in plants

The organic compounds such as carbohydrates, proteins, fats, lipids, nucleic acid, chlorophylls are primary metabolites.

Secondary metabolites are secondary plant products formed during the metabolic activity of the plants. The secondary metabolites may be grouped into three,

- Isoprenoid compounds or terpenes. Example: essential oils, steroids, rubber etc.,
- Nitrogen containing secondary metabolites. Example : alkaloids, non protein amino acid etc.,
- Phenolic compounds or phenolics like lignin, tannin, flavonoids.

Isoprenoids or terpenes show properties of lipids and are synthesized from acetyl CoA through mevalonic acid pathway.

Phenolics and aromatic compounds are synthesized in two ways either from acetyl CoA via Mevalonic acid pathway or from erythrose 4 phosphate and PEP via shikimic acid pathway

Nitrogen containing secondary metabolites such as alkaloids are synthesized in plants primarily from, amino acids.

Ligand efficiency

It is a measurement of binding energy per atom of a ligand to its binding partner, such as receptor or enzyme.

K_m for an enzyme

Experimentally, the Km for an enzyme tends to be similar to the cellular concentration of its substrate. An enzyme that acts on a substrate present at a very low concentration in the cell usually has a lower K_m than an enzyme that acts on a substrate that is more abundant.

Pyrethroid

Pyrethroids, a monoterpene ester is found in the leaves and flower of Chrysanthemum species, show insecticidal activity.

A protein in 100mM KCl solution was heated and the observed T_m (midpoint of unfolding) was 60°c. When the same protein solution in 500mM KCl was heated the observed Tm was 65°c. The reason for the increase in T_m?

The temperature which disturbed the structure of protein is called T_m (melting temperature). When the protein is placed in highly concentrated solution, more hydrophobic interaction and electrostatic repulsion are produced due to which the melting temperature is increased.

An amino acid contains no ionisable group in its side chain (R). It is titrated from pH 0-14. Ionisable states that were observed during such titration are,

Acid base titration involves the gradual addition or removal of protons. At a very low pH the predominant ionic species of glycine is 'H_3N-CH_2-COOH'. At the midpoint equimolar concentrations the proton donor (^+H_3N-CH_2-COOH) and proton acceptor (^+H_3N-CH_2-Coo$^-$) species are present.

An α helix in a peptide or protein is characterized by hydrogen bonds and characteristic dihedral angles:

The simplest arrangement of polypeptide chain could assume with its rigid peptide bonds is a helical structure called α helix. The amino acid residues in an α helix have conformation with Ψ= -45° to -50° and Φ= -60° and each helical turn includes 36 amino acid residues. The helical twist of α helix found in all proteins is right handed. The structure is stabilized by a hydrogen bond between hydrogen atoms attached to electronegative nitrogen atom of the peptide linkage and electronegative. Carbonyl oxygenation of the fourth amino acid on the amino terminal side of that peptide bond.

The stereocilia of auditory hair cells are arranged in rows but the heights of stereocilia are not the same in all the rows. Through the height of stereocilia is the same within a particular row, the height increases in subsequent rows. When the stereocilia of shorter rows are mechanically pushed towards the taller rows, the hair cells are depolarized but a push on opposite direction hyperpolarizes them. The significance of this graded height in stereocilia is the tip of the taller stereocilia will show greater displacement as compared to shorter once when all the rows are moving in the same axis. The hair cells will be depolarized or hyperpolarized in different grades when the axis of displacement is changed

Each hair cell contains a bundle of stereocilia and a loan kinocilium. Across the epithelium, hair cell polarizations are topographically organized, with an imaginary line running through the epithelium about which stereocilia are oppositely directed. The mechanism responsible for morphological polarization (also termed planar cell polarity) has been well studied, particularly in auditory hair cells. Stereocilia consists of actin filaments within a tubular membrane. Because the filaments are cross bridged, each stereocilia behaves like a stiff rod which pivots at its base. The adequate stimulus for the hair cell is displacement of the hair bundle. The hair cell is a mechanoreceptor, producing an electrical signal or receptor potential, in response to mechanical stimulation of its hair bundle. It is the relative motion between the hair cell and the auxiliary structure specific to the receptor organ that provides the displacement. The change in membrane potential associated with the movement of the hair bundles results in changes in the discharges of 80 nerve afferent axons connected at the hair cell base. Depolarization of the hair cell leads to increased firing of the fibre, while hyperpolarisation results in cessation of firing. This relationship between the electrical properties of hair cells and their 80 nerve discharge patterns is referred to as functional polarization.

Phenyl alanine ammonia lyase (PAL) *and chalcone synthase* (CHS) *are involved in biosynthesis of phenolic compounds in plants.*

Theory: The plant phenolic compounds are primarily derived from the phenylpropanoid pathway and acetate pathway. The phenyl propanoid are products of the shkimic acid pathway (phe, tyr). Phenylalanine ammonia lyase (PAL) and CHS are the two key enzymes that are involved in phenyl propanoid biosynthesis the first reaction of this biosynthetic pathway is the deamination of phenylalanine to cinnamic acid by PAL. In the next reaction of the metabolic pathway CHS condenses 3 malonyl co-A molecules with cinnamoyl CoA to produce chalcone. This condensation is the main branch in the pathway for the production of flavonoids. PAL and CHS catalyze key reactions in the biosynthesis of phytoalexin isoflavonoids in legumes. This is the first stage in the biosynthesis of secondary phenylpropanoid products by L-phenylalanine and the first reaction branch in the main flavonoid and isoflavonoid production pathway

A plot of V/[S] versus V is generated for an enzyme catalysed reaction and a straight line is obtained, the information that can be obtained from the plot are Vmax, Km *and turnover number.*

Reason: The Km value for an enzyme depends on the particular substrate and environmental conditions such as pH, temperature, ionic strength etc., the maximum rate Vmax reveals the turnover number of an enzyme which is the number of substrate molecules converted into products by an enzyme molecule in a unit time when the enzyme is fully saturated with substrate. A plot of 1/V at Y as a function of 1/[S] at X gives a straight line

Enzymes are nowadays used extensively in bioprocessing industries. Enzyme 1 is used for treatment of hides to provide a finer texture in leather processing and manufacture of glue. Enzyme 2 is used for classification of fruit juices. The enzyme 1 is protease and enzyme 2 is pectinase.

Reason: The cloudiness of fruit juices and Vines is due to pectins. The pectin is digested by proteolytic enzymes prepared from

Aspergillus niger. These enzymes preparations are mixture of polygalacturonase, pectin esterase, pectin lyase and hemicelluloses.

The bacterial flagellar motor is a multi protein complex. Rotation of the flagellum requires movement of protons across the membrane facilitated by a multi protein complex. The flagellar motor proteins combine to create a proton channel that drives mechanical rotation. In a screen for mutants, some non motile ones were selected. These could have mutated H^+-ATPase.

Reason: Bacterial flagellum is composed of three basic parts the filament, hook and basal body. The movement of prokaryotic flagellum results from rotation of its basal body. The exact mechanochemical basis is not known but believed that it depends on cell's continuous generation of energy (mutated H+-ATPase lacks movement of flagellum).

The lifetime of a peptide bond in protein is very large (1000 years).

The stability of peptide is due to the energy barrier to be crowd to go to the hydrolyzed state is large. Peptide bond hydrolysis is favourable and means free energy always negative. Peptide bond easily hydrolyzed in presence of 6N HCl at 100 °C.

Two homologous proteins were isolated from a psychrophile (P) *and a thermophile* (T) *the purified proteins were subjected to denaturation, protease digestion and circular dichorism* (CD). *Following observations were made*

- *The* CD *spectra of* P *and* T *proteins are identical*
- *Their amino acid composition is 95% identical*
- *T and P are equally susceptible to proteolysis in the presence or absence of seducing agent.*
- *T has higher midpoint of thermal denaturation than* P

Reason for enhanced stability in T is due to increase number of disulphides in T. The presence of disulphide bonds makes protein more stable at high temperature.

CONCEPTS RELATED TO PHYSIOLOGY

cAMP dependent pathway

It is also called adenylyl cyclise pathway (G- protein coupled receptors). These GPCRs binds to and is activated by specific ligand stimulus that ranges in size from small molecule catecholamines, lipids or neurotransmitters to large protein hormone. Gs alpha subunits of G protein binds to and activates an enzyme called adenylyl cyclase. It catalyzes the conversion of ATP to cAMP (Adenosine Monophosphate). Increase in concentration of cAMP lead to activation of

- cyclic nucleotide gated ion channels
- Exchange proteins activated by cAMP (EPAC) such a RAPGEF 3
- Popeye domain containing proteins (popde)
- an enzyme called Protein kinase A (PKA)

PKA enzyme is also known as cAMP dependent enzyme because it gets activated only if cAMP is present. Once PKA is activated, it phoshorylates a number of other protein including:

- Enzymes that convert glycogen into glucose.
- Enzymes that promote muscle contraction in the heart leading to an increase in Heart rate.
- Transcription factors which regulate gene expression also phosphorylates AMP.

Function:

Increases in heart rate, cortisol secretion and breakdown of glycogen and fat. cAMP is for the maintenance of memory in the brain, relaxation of the heat and water absorbed by the kidney. This pathway can activate enzyme and regulate gene expression.

If cAMP dependent pathway is not controlled, it can ultimately lead to hyper proliferation, which may lead to cancer.

Molecules that activate cAMP pathway:

- Cholera toxin - increases cAMP levels.
- Forskolin - a diterpene natural product that activates adenylyl cyclase.
- Caffeine and theophylline - inhibit cAMP phosphodiesterase, which degrades cAMP thus enabling higher levels of cAMP than would otherwise be had.
- Bucladesine (dibutyryl cAMP, dbc AMP) - also a phosphodiesterase inhibitor.
- Pertussis toxin- increases cAMP levels by inhibiting Gi to its GDP (inactivate) form. This leads to an increase in adenylyl cyclase activity, thereby increasing cAMP levels, which can lead to an increase in insulin and therefore hypoglycaemia.

Deactivation:

Gs alpha submit slowly catalyzes the hydrolysis of GTP to GDP, which in turn deactivates the Gs protein, shutting off the cAMP pathway. The pathway may also be deactivated downstream by directly inhibiting adenylyl cyclase or dephosphorylating the protein phosphorylated by PKs.

Molecules that inhibit cAMP pathway include:

- cAMP phosphodiesterase dephosphorylate cAMP into AMP reducing the cAMP levels.
- Gs protein which is a G protein that inhibit adenylyl cyclases thus reducing cAMP levels.

Adenylyl cyclase

- It is the enzyme with key regulatory roles in all cells.
- Catalyse the conversion of ATP to cAMP and pyrophosphate.

- Mg ions are generally required and appear to be closely involved in enzymatic mechanism.

Classes of AC enzymes

Class I enzymes:

- Occur in Bacteria.
- E.coli deprived of glucose produce cAMP that serves as an internal signal to activate the expression of gene for importing and metabolizing other sugars.

Class II enzyme:

- Toxins secreted by pathogenic bacteria like *Bacillus anthraces* during infection
- Example: cya A

Class III enzymes

- Occur in *Mycobacterium tuberculosis* and play a key role in pathogenesis.
- These are integral membrane proteins involved in transducing extracellular signals into intracellular responses.
- In human liver adrenaline indirectly stimulates AC to mobilize stored energy in the fight or flight response. This happens through G protein signalling cascade.
- cAMP is an important molecule in eukaryotic transduction hence referred to as second messenger.
- Photoactivable adenylyl cyclase (PAC) play role in neural activity and behaviour of organisms.
- Example: certain neurons have been shown to alter the glooming behaviour in fruit flies exposed to blue light.
- Adenylyl cyclase is implicated in memory formation where it functions as a coincidence detector.

Class IV enzymes:

> It is the smallest of AC enzyme classes.

Intracellular as well as extracellular cAMP plays a crucial role in gene regulation during development. During aggregation, cAMP pulses strongly accelerate the expression of components of cAMP signalling system. During post aggregative development cAMP directly induces entry into the spore differentiation pathway and by inducing the synthesis of stalk cell inducing factor. DIF, cAMP is also indirectly responsible for the differentiation of stalk cells. cAMP produced by three structurally distinct ACs (adenylyl cyclases), ACA, ACG, ACB which have distinct but overlapping patterns of expression and as concluded from gene disruption studies, seemingly overlapping functions. In addition to gene disruption, acute pharmacological abrogation of protein activity can be a powerful tool to identify the protein's role in the biology of the organism. In *Dicotyostelium,* cells express a G-protein coupled adenylyl cyclases, ACA during aggregation and a typical adenylyl cyclase ACG in mature spores. The ACG gene was disrupted by homologous recombination, acg-cells developed into normal fruiting bodies with variable spores, but spore germination was no longer inhibited by high osmolarity, a fairly universal constraint for spore and seed germination. ACG activity, measured in aca-/ACG cells, was strongly stimulated by high osmolarity with optimal stimulation occurring at 200 milliosmolar. Red C mutants, which display unrestrained protein kinase A (PKA) activity and a cell line, which over expresses PKA under a pre spore specific promoter, germinate very poorly, both at high and low osmolarity. There data indicate that ACG controls spore germination through activation of PKA

Pathways controlling flowering time in Arabidopsis

Arabidopsis is a long day plant, which flowers earlier under long days but eventually flowers under short days.

The genes CONSTANS(Co), CRYPTOCHROME E2/FHA(CRY2), GIGANTEA (GI), FT and FWA are part of long day promoting pathway.

- FI and FWA act downstream of Co
- GI and CRY2 act upstream of Co
- Co – long day promoter.

Co – encode protein with two zinc fingers loosely related to those of GATA transcription factors and contain carboxyl terminal domain called CCT

> *Zinc finger is similar to B-box and a type of zinc finger identified in animal proteins are believed to mediate protein-protein interactions.*

Mutation of Co results in late flowering

Arabidopsis is a long day plant, in order to induce late flowering, the amount of the gene Constans should be more whose mutation suppress flowering and in short day plants, the concentration of Constans is lower since it is a long day promoter.

Sucrose uptake in plants

Plants are autotrophic organisms that are able to synthesis complex molecules by reducing C, N and S from simple molecules. As a major translocatable product of photosynthesis, sucrose (glucose+fructose) is the major soluble component of the phloem sap.

Selection of sucrose as the highly transported sugar in plants has been related to its non-reducing nature and relative insensitivity to metabolism. This represents an advantage for a substrate translocated over long distance in plants allowing transport without the problem of metabolism easily encountered with glucose.

In plants, sucrose is transported over long distance in solution in the phloem sap. This flow of sap occurs in sieve elements. Sieve elements are connected to form sieve tubes that oppose very little resistance to the flow of sap. The movement of sap in the phloem occurs through mass flow; the driving force for the movement being the entry of sucrose and subsequently water in the sieve tubes in the source organ while at the other end of the conduit in the sink organs, the continuous uploading of solutes and water would maintain the flow.

The accumulation of sucrose in sieve tube requires the presence of sucrose transporters to drive this active accumulation. This points to the importance of this carrier system for the translocation of solutes from sucrose to sink organs. The existence of a carrier system specific for sucrose and responsible for the entry of sucrose in the phloem.

The energy of this transport being the phloem gradient established by a H^+/ATPase located in plasma membrane.

In short, at lower concentration of sucrose, the uptake of sucrose is energy dependent and carrier mediated while on increasing concentration it becomes energy independent.

Atmospheric Co_2 assimilation by C_3 and C_4 plants

Atmospheric Co_2 contains the naturally occurring stable carbon isotope 12C and 13C in the proportion of 98.9% and 1.1% respectively. The dark reaction of photosynthesis involves Co_2 assimilation to form carbohydrate. Both C_3 and C_4 plants assimilate lesser $13Co_2$ to $12Co_2$. C_3 plants assimilate lesser $13Co_2$ as compared to C_4 because C_3 plants are less efficient to Co_2 fixation as compared to C_4 plants. Isotopic technique revealed that C_4 plants (Corn) have 13C: 12C ratio referred as δ13C values ranging between -8 to – 13% whereas C_3 plants (nectar bearing plants) have values between -22 and -30%

Co_2 assimilation in C_3 plants

- Found in all photosynthetic plants
- It includes hydrophytes, mesophytes and xerophytes.

- The plants have photoactive stomata
- High rate of photorespiration is present in these plants.
- The anatomy of the leaf is normal
- For the synthesis of one molecule of glucose 12 NADPH and 18 ATPs are needed.
- There is a single CO_2 fixation
- Primary acceptor of atmospheric carbon dioxide is RUBP
- The first stable compound formed is 3 phosphoglyceric acid.
- The first enzyme involved in CO_2 fixation process is RUBISCO
- The CO_2 compensation point is 30 – 70 ppm.

CO_2 assimilation in C_4 plants

- It includes only tropical plants viz., mesophytes.
- The plants have photoactive stomata.
- The rate of photorespiration is less or negligible.
- The anatomy of leaf is a special type called Kranz anatomy.
- For synthesis of one molecule of glucose 12 NADPH and 30 ATPs are needed.
- Double CO_2 fixation occurs.
- The acceptor of atmospheric CO_2 is PEP (present in mesophyll cell) and the metabolic CO_2 acceptor is RUBP (present in bundle sheath cell).
- The first stable product formed is Oxaloacetic acid(OAA)
- The first enzyme involved is PEPcarboxylase.
- The carbondioxide compensation point is 10ppm

Overview of C3, C4 and CAM plants

C_3 plants

The first molecule formed in the fixation of carbon dioxide is a 3 carbon compound (3PGA). 85% of the plants use C_3 pathway to fix carbon via calvin cycle. Enzyme RUBISCO cause oxidation reaction in which some of the energy used in photosynthesis is lost in a process known as photorespiration. 25% reduction in the amount of carbon is fixed by plant and released back to the atmosphere as carbon dioxide.

CAM and C_4 plants have added steps to help concentrate and reduce the loss of carbon during the process.

C_4 plants

The carbon dioxide fixation pathway is termed Hatch Slack pathway in which a 4 carbon compound malic acid or aspartic acid is formed as an intermediate. Co_2 that is taken in is moved to bundle sheath cell by malic acid or aspartate. O_2 in bundle sheath is very low, so RUBISCO enzymes are less likely to catalyze oxidation reactions and waste carbon molecules.

Crassulacean Acid Metabolism (CAM)

CAM plants are succulents that are efficient in storing water. In these plants stomata are closed during day. Co_2 is taken in at night and converted to a molecule called malate which is stored until the daylight return and photosynthesis begins via Calvin cycle.

Menstruation

Menstruation is bleeding from the uterus of adult female at interval of one lunar month (28 days) on an average. The menstrual cycle consists of proliferative phase (14 days), secretory phase (10 days) and menstrual phase (4 days). The follicle stimulatory hormone (FSH) secreted by anterior pituitary stimulates ovarian follicle to secrete oestrogen. Oestrogen stimulates the proliferation of endometrium of the uterine wall. The luteinizing hormone (LH) secreted by anterior pituitary causes ovulation. The initial FSH level should be higher than LH for menstrual cycle.

Hormones in female reproductive cycle

- ➤ Day one of menstruation- estrogen and progesterone levels are low
- ➤ This signals the pituitary to produce FSH resulting in the maturation of follicle.
- ➤ The follicle produces more estrogen for ovulation
- ➤ After 12-14 days of ovulation increased estrogen trigger sharp rise in LH which cause the release of egg from the follicle.
- ➤ Ruptured corpus luteum secretes progesterone.

Aquaporins

Aquaporins is a family of integral proteins which provide channels for rapid movement of water molecules across the plasma membrane in a variety of specialized tissues. In both plants and animal cell membranes aquaporins are present. Most aquaporins does not allow passage of ions or other small solutes. All aquaporins are typeIII integral proteins with 6 transmembrane helical segments.

Photoreceptors in model plant Arabidopsis thaliana

Arabidopsis have five different phytochromes viz., Phy A, B, C, D, E

Seedlings containing mutations in Phy A or phyB gene grew taller than the wild plant under continuous far red or red light respectively, demonstrating the function of Phy A and Phy B in the perception of the corresponding wavelength of length for the hypocotyls inhibition response. Over expression of PhyA results in photoperiod insensitive early flowering under day extension conditions. Phy A mutant flowered significantly later than the wild type. Phy C is involved in red light induced hypocotyls inhibition and leaf expansion. PhyC is the regulatory photoreceptor that exists in two forms that are reversibly inter convertible by light; the pr form that absorbs maximally in the red region of the spectrum and pfr absorb maximally in the far red region. The LOV (Light oxygen voltage) domain of Phy C is an important domain for signal transmission. Phy D and E act in conjugation with Phy B in the regulation of much shade avoidance responses.

Arabidopsis also has multiple blue light receptors including Cryptochromes such as Cry 1 and Cry2 and a protein kinase NPH1. Cry 1 mediates blue light induced inhibition of hypocotyls elongation and anthocyanin biosynthesis. It is the major blue light receptor mediating blue light inhibition of hypocotyls elongation. Cry2 regulates floral inhibition in response to photoperiod. Cry1 and Cry2 positively regulate floral initiation. Cry1 caused flowering in short day plants and Cry2 caused flowering in long day plants. NPH 1 is a flavoprotein possessing a light regulated kinase activated responsible for the blue light dependent phototropism.

Statements on low temperature stress

Membrane lipids isolated from chilling resistant plants often have a greater proportion of unsaturated fatty acids than those from chilling sensitive plants. During acclimation to low temperature the activity of disaturase enzymes increases and the proportion of unsaturated lipid rise. This modification lowers the temperature at which the membrane lipids begin a gradual c phase change from fluid to semi crystalline and allows membranes to remain fluid at lower temperatures. Thus desaturation of fatty acids provides some protection against damage from chilling.

During episodes of anoxia in plants, pyruvate produced in glycolysis is initially fermented to lactate. During later stage, there is an increase in the fermentation to ethanol and decrease in the fermentation to lactate, a phenomenon which helps plants survive anoxia. This is because the cytosolic pH decreases, thus inhibiting lactate dehydrogenase and activating pyruvate decarboxylase activity.

Reason: The condition of absence of oxygen is called anoxia. Glycolysis yield a net profit of 2 ATP. Two $NADH^+H^+$ and glucose was split to 2 pyruvate molecules. In the absence of oxygen, pyruvate undergoes fermentation during which $NADH^+H^+$ is converted back to NAD^+

Fermentation can be of two types,

Alcohol fermentation

Pyruvate is decarboxylated (CO_2 is released) to form acetaldehyde. Hydrogen atom from $NADH^+H^+$ is then used to convert acetaldehyde to ethanol. This is catalysed by the enzyme pyruvate decarboxylase.

Lactate fermentation

Two molecules of Pyruvate are converted to two lactate molecules, which ionise to form lactate.

$NADH+H^+ \longrightarrow NAD$

This reaction catalyzed by the enzyme Lactate dehydrogenase.

There will be a fall in cytoplasmic (cyt) pH followed by the onset of anoxia.

Spinal cord of an animal was transacted at the C1/C2 level. The respiration of the animal stopped and is needed artificial respiration but the heart continued to beat although at a slower rate because heart has auto regulation.

Regulation of Heart beat

Cardiovascular centre is the part of the human brain responsible for the regulation of rate at which the heart beats through the nervous and endocrine systems. It is located in medulla oblongata. Normally heart beats without nervous control, but in some situations (like exercise, body trauma) the cardiovascular centre is responsible for altering the rate at which the heart beats. It also mediates respiratory sinus arrhythmia.

The autoregulation of heart is classified into two types:

Homeometric autoregulation

It is the heart's ability to increase contractility and restore stroke volume when after load increases. This is in contrast to heterometric regulation.

Heterometric autoregulation

It is the intrinsic regulation of the strength of cardiac contraction as a function of diastolic fibre length (volume) independent of the autonomous nerves and other influences.

Systole and Diastole

- Diastole occurs when the mitral valve (MV) open so that the left atrial (LA) and left ventricle (LV) pressures are equal. Late Diastole results when a small rise in pressure in both LA and LV.
- Systolic contraction occurs when the pressure of LV rises and if it exceeds LA pressure, the MV closes, contributing to the first heart sound (S 1).
- As LV pressure raises above aortic pressure the aortic valve (AV) opens which is a silent event. As ventricle begins to relax and its pressure falls below the aorta, the AV closes contributing to second heart sound (S2).
- Norepinephrine produce decreased heart rate and increased blood pressure in dogs.

Cardiac cycle

Cardiac cycle is the sequence of events related to flow of blood pressure that occurs during one heart beat. It is composed of atrial systole and diastole (both of a total 0.8s) and ventricular systole and diastole (both of total 0.8s). From this we could calculate,

$t_{as} + t_{ad} = 0.8$

$t_{vs} + t_{vd} = 0.8$

$t_{as} + t_{ad} = t_{vs} + t_{vd}$

For a normal heart, the time taken for atrial systole and diastole are A_s and A_d seconds respectively, while the same for ventricular systole and diastole are V_s and V_d. Correct equation is

$A_s + A_d = V_s + V_d$

Theory: The action of heart includes contraction and relaxation of atria and ventricles. The contraction of heart is called systole while relaxation is called diastole and collectively called heart beat. The time of one heart beat is 1.6s. It shows that

$A_s(0.1) + A_d(0.7) = V_s(0.3) + V_d(0.5)$

An organism having heart for circulation excretes through green glands. It has several ganglia and tactile organs on its body and its larval form is very different than its adult form. This organism is most likely to respire by gaseous exchange over thinner areas of cuticle or by gills.

Reason: The excretory organs of some crustaceans (Crayfish and crabs) are antennal glands or green glands. In those crustaceans that have gills for respiration nitrogenous wastes are removed by simple diffusion across the gills. Oxygen and Carbon dioxide are exchanged between blood and water across the gill surface.

Crustaceans have open circulatory system, eyes on stalks, primitive ventral cord and brain (ganglia neat the antenna), a digestive system which is a straight tube for grinding food and a pair of digestive glands. Gills are used for respiration and they have a pair of green glands to excrete waste.

Excretory organs of Arthropods

Nephridia: Found in periplates, it is situated on lateral side of segmented body cavity.

Coxal glands: Characteristics of Arachnida

Green glands: It is also called antennal gland and is found in Malacostraca and larval forms of crustaceans.

Shell glands: it is also called maxillary glands. Found in Branchiopoda, Ostracoda, Lopepoda, Cirripedia and some larvae of Crustaceans.

Malphigian tubules: These are filamentous bodies with or without lumen made of ciliate or cubical epithelium. It is found in all insects except Collembola.

Hepatopancreas: Found in Limulus. They shed large amount of calcium phosphate as excretory product.

Fatbody: Found in certain insects like Myriapoda and Onychophora. The fat body is made up of polygonal cells. As they grow old, gets filled with minute urate crystals.

Exoskeleton: Found in Crustacea and Insects. The nitrogenous substances secreted by these organisms remain deposited within theexoskeleton.

Intestinal caeca: Found in Squilla

Midgut epithelium: Found in Naupliius larvae of Crustacea.

Pericardial cells: In insects some cells around the heart and pericardial membrane are excretory in function.

Nephocytes: They are found within Haemocoel of insects.

Oenocytes: These are found around abdominal spiracles in Insects and Myriapoda.

GnRH is secreted during infancy (0-6 months) and puberty onwards (4 years and above) in monkeys. However i.v. injection of GnRH during pre-published period (about 2 years of age) lead to elevated Lh and FSH in blood compared to untreated 2 years old monkey because pituitary is active in all the stage of development in monkey.

Reason: Gonadotropin-releasing hormone (GnRH) is secreted from hypothalamus. It stimulates the anterior lobe of pituitary to secrete FSH and LH. High levels of LH and FSH in after injection of GnRH states that pituitary is active at all stages.

The intestinal absorption of glucose is impaired by the use of quabain, an inhibitor of Na^+/K^+ ATPase is because the inhibitor has blocked Na^+ Transport from epithelial cells to the intestinal space.

Reason: Glucose and galactose are absorbed by active transport. Na^+ pump of the cell membrane helps in its active take up. Glucose and galactose are absorbed into the blood capillaries. Quabain is an inhibitor which blocked Na^+ transport from epithelial cells to intestinal space.

Water potential (Φ) reduction and cellular dehydrations

While studying the primary effects of different abiotic stresses on plants, a researcher observed water potential (Φ) reduction and cellular dehydrations, a combination of freezing, salinity and water deficit stresses could have caused the observed effect.

Red and for red light are perceived by plants through various photoreceptors including phytochromes. The activation of phytochrome is caused by the conversion of pr to pfr form through the effect of red light.

Theory: Phytochrome is an amorphous photoreceptor chromoprotein on protein pigment which exist in two states pr and pfr. In higher plants, it occurs in most of the organs including flowers, fruits and seeds. The pr form is cytosol associated while pfr form is membrane associated

Upon absorption of a photon, a chlorophyll molecule gets converted to its excited state when the energy of the photon is equal to the energy gap between ground state energy and the excited energy state.

Theory: In photosynthesis, light harvesting molecular are of two types' antenna molecule and core molecules. On absorption of light energy, the antenna molecules get excited. The excited antenna molecules hand over their energy to core molecules by resonance and come to the ground state. The energy picked up by core molecules is supplied to the trap or photo centre. The energy of the photon is equal to the energy gap between ground state energy and excited state energy

The localisation of photosynthetic supramolecular complexes on plastic lamellae can be easily understood with the statements,

- PSII *is preferentially located on granal lamellae.*
- ATP *synthase and* PS1 *are preferentially located on stroma lamellae.*

Reason: Photosystem II is located in appressed part of grana thylakoids while PS 1 is located in non appressed part of grana thylakoids and stroma lamellae. Cytochrome b6/f complex is a membrane bound complex associated with non cyclic photophosphorylation

In a tissue, cells are bound together by physical attachment between cell to cell or between cells to extracellular matrix. Adherens junctions are cell to cell anchoring junctions connecting actin filament in one cell with that in the next.

Desmosomes or cell Matrix anchoring junctions connecting actin filament in one cell to extracellular Matrix.

Reason: The light junction sometimes called terminal bar or sheet like junctions that connect one cell to another without any communication between the cells and containing sealing stands. Gap junctions permit controlled passage of small molecules or ions between the cells. Desmosomes are mechanical junctions which attaches cytoskeleton of one cell to the cytoskeleton of the other cell or to the extracellular matrix.

Level of follicle stimulating hormone (FSH) during infancy and adulthood is the same but spermatogenesis is seen only during adulthood. mRNA levels coding for FSH receptor are also found to be the same in testis of both age groups. The investigations that will clarify this paradox a little more is culturing the testicular cells and add FSH to see comparative rise in cAMP production by both age groups.

Reason: FSH stimulates growth of ovarian follicles and secretion of oestrogen in female and spermatogenesis in male. Interstitial cell stimulating hormone (ICSH) activates the Leydig's cell of testis to secrete androgen. In female(here called LH) it stimulates the Corpus luteum of the ovary to secrete progesterone that is, the comparison in CMP production will clarify this paradox in both age groups.

The microorganisms involved in the global nitrogen cycle are Rhizobium, Nitrosomonas, Nitrobacter, Pseudomonas, Azotobacter.

Theory: Denitrication is a process of nitrate reduction that ultimately produce molecular nitrogen. It involves *Pseudomonas*. Nitrogen fixation is the conversion of atmospheric nitrogen to utilizable compounds like nitrate, ammonia and amino acids. It involves *Rhizobium* (symbiotic), *Azotobacter* (free living) etc., Nitrification is conversion of ammonium nitrogen to nitrate nitrogen with the help of *Nitrosomonas* and *Nitrobacter*.

One of the important functions of programmed cell death (PCD) in plants is protection against pathogens. PCD also appears to occur during the differentiation of xylem tracheary elements that lead to nuclei and chromatin degradation. These changes results from the activation of certain genes. The genes involved in the differentiation of xylem tracheary elements are Topoisomerase and protease.

Reason: Extracellular proteolysis plays a key regulatory role during PCD. The secondary cell wall synthesis and cell death are coordinated by c secretion of 40KD protease and secondary wall precursors. DNA topoisomerase is involved in the induction of Apoptosis like PCD (AL-PCD). Secondary wall synthesis is accomplished by synthesis of nuclease and proteases, vacuolization of the cytoplasm and the influx of Ca_2^+

If an Arabidopsis plant, mutated in lycopene biosynthetic pathway is grown in sunny tropical climate in the presence of oxygen it would show reduced biomass due to photo oxidative damage.

Reason: In the lycopene biosynthetic pathway the final product all Trans lycopene synthesized in the final step in this pathway is the founder precursor for carotenoid biosynthesis in plants. The light dependent generation of active oxygen species is termed as photooxidative stress. Photo inhibition and photo oxidation only occur when plants are exposed to stress.

AP1 (APETALA-1) is one of the floral meristem identifying genes. In wild type Arabidopsis thaliana plants transformed with AP1::GUS, glucuronidase (GUS) activity is seen in floral meristem, only after the commitment to flowering ectopic expression of AP1::GUS in the EMBRYONIC FLOWER (emf) mutant background results in GUS activity throughout the shoots in four day old seedlings. These observations suggest that AP1 stimulates the flowering in the emf background.

Reason: *Arabidopsis* have at least three domain genes have been isolated affect the timing of flower formation. At least five genes are known that impart identity on floral meristem, when mutated these genes result in either shoots instead of flowers or in highly abnormal flowers. These genes include leafy, unusual floral organs, Apetala 1 and Apetala 2. The GUS activity throughout the shoot after ectopic expression of AP1::GUS in emf mutant background is due to AP1 stimulation in emf background.

A monkey undergoes cerebellectomy. After the post operative recovery, the monkey was given a task to press a bar. It was observed that, its hand would overshoot the target while reaching the bar. The monkey exhibited intention tremor while trying to press the bar.

Theory: Cerebellum plays a role in eye movement, just as in other movement. This is because,

- Disease of the cerebellum causes a variety of oculomotor defects.
- Stimulation of the cerebellum causes eye movement.
- Electric potentials appear in the cerebellum when the eyes move.
- Intimate anatomic connections link cerebellum with mid brain and pontine regions immediately concerned with eye movement. One way of investigating the role of the cerebellum is to observe defects in the primates with complete cerebellectomy. If nothing else it will show what functions are spared for this cerebellum is not indispensable. The cerebellum processes input from other areas of the brain, spinal cord and sensory receptors to provide precise timing for coordinated, smooth movements of the skeletal muscular system.

CONCEPTS RELATED TO IMMUNOLOGY

Transplant rejection

The replacement of diseased tissue or organs by healthy one is called transplantation. Sometimes transplantation may be rejected by the recipient body. When a person receives an organ from someone else the person's immune system may recognize that as foreign. Doctors use medicine to suppress the recipient's immune system. Cornea transplants are rarely rejected because the cornea has no blood supply. The transplant between identical twins is never rejected.

Types of rejection

There are three types of rejection,

Hyperacute rejection

Occur few minutes after the transplant when the antigens are completely unmatched. The tissue must be removed away to avoid the death of the recipient. This is mostly seen in transfusing mismatch blood type.

Acute rejection

Takes place from first week after transplant to three months after transplantation.

Chronic rejection

The rejection takes place and remain prolonged for many years.

Methods to prevent graft rejection

To prevent graft rejection following methods are used,

- ➢ Tissue matching
- ➢ Exposure of bone marrow and lymph tissue to radiation

- Immunosuppressors
- Treatment with antibiotic reducing infection

Fluorescent Activated cell sorter (FACS)

Fluorescent Activated cell sorter (FACS) can be used for very rapid (upto 1000 cells/sec) sorting of transformed cells. This is applicable to the entire genus. Whole product becomes arranged on the cell surface and available for binding for specific antibodies. The protein product of DNA insert can be identified by the unique function viz., a function not performed by the proteins of non transformed host cells. The assay for green fluorescent protein can be performed by exposing the intact tissue to ultraviolet light which produces a green fluorescence. Thus it had added advantages over the rest because the expression of the promoter can be studied in intact tissues and cells.

Erythropoietin

During the Spanish conquest of the Inca empire at the high altitude in Peru, many soldiers fell sick. It was found that the sickness was due to low partial pressure of O_2 in the atmosphere at that altitude. To determine the reason, blood was collected from those patients. Erythropoietin (EPO) is a glycoprotein hormone that controls production of erythrocytes, the process called erythropoiesis. The native population has high level of EPO thus live normally. While the soldiers had low EPO level and thus less haemoglobin saturation at low oxygen tension.

Mouse bone marrow cells were fractionated to derive stem cell antigen - 1+[sca-1+] cells. These cells were cultured with interleukin-3 and granulocytes- macrophage colony stimulating factor, or macrophage colony stimulating factor or granulocyte colony stimulating factor. But most numerous and varied colonies were obtained in the culture stimulated with Macrophage stimulating factor.

This is because; the molecular mechanisms of constitutive macrophage distribution and induced migration involve cellular adhesion molecules, cytokinins and growth factor as well as chemokines and chemokine receptor. Macrophage colony stimulating factor (MCSF) is a major growth differentiation and survival factor selective for macrophages, whereas Granulocyte macrophage colony stimulating factor (GMCSF) regulates myeloid cell production and function.

Cancer causing genes can be functionally classified into mainly three types (a) genes that induce cellular proliferation (b) tumor suppressor genes (c) genes that regulate apoptotic pathway. Epstein Barr virus that cause cancer by modulating apoptotic pathway, contain a gene having sequence homology with the gene bcl-2.

Reason: There are two mutational routes towards the uncontrolled cell proliferation and invasiveness that are characteristics of cancer. The first is to make a stimulatory gene hyperactive viz., proto-oncogene to oncogene alteration and second is to make an inhibitory gene inactive called tumor suppressor gene. In apoptotic pathway bcl-2 gene inhibits programmed cell death and thus act as an oncogene.

Toll like receptor 4 is associated with responsiveness to LPs *an endotoxin that causes lethal endotoxic shock. The mice deficient in toll like receptor 4 and* BALB/c *mice were injected with E.coli. In addition, some* BALB/b *mice were also injected with the same bacteria alone or with anti interleukin-10*(IL-10) *antibody. The mice resistant to lethal bacteria were the mice deficient in toll like receptor.*

Reason: Many bacterial PAMPs (Pathogen associated molecular patterns) activate cells via Toll like receptors (TLR). Therefore absence of toll-like receptor causes resistancy to the lethal effect of bacteria.

Intracellular pathogens like Mycobacteria, Salmonella, Leishmania and Listeria survive in macrophages by modulating host cellular machinery. Inorder to study the fate of these intracellular pathogens in macrophages, cells were labelled with Lysotracker Red and infected with GFP labelled organisms. After 2 hours at 37° C cells are fixed, stained with anti transferrin receptor antibody and probed with secondary antibody conjugated blue dyes. Cells were viewed under confocal microscope

Observation: GFP labelled Mycobacteria, Salmonella and Listeria were localized in the same compartment labelled with blue dyes, whereas GFP Leishmania co-localize with red labelled compartment because, Leishmania resided in lysosome like compartment.

Reason: In the absence of opsonins (molecules such as antibody, complement and C-reactive protein that promote phagocytosis) macrophages uses multiple receptors to recognize and engulf a wide range of microorganisms. Different receptors often collaborate, for example CR3 and MR in the uptake of *Leishmania promastigotis.* Non-opsonic uptake or invasion results in a range of survival strategies by intracellular pathogens enabling them to avoid phagosome maturation, inhibit fusion with lysosome and acidification (as for *Mycobacterium*). *Salmonella* induces formation of spacious phagosome while *Leishmania* replicate in secondary lysosome within vacuole.

Macrophages were collected from BALB/c *mice,* CD40 *deficient mice. These macrophages were co cultured with* LCMV *peptide specific T cells in the presence of the* LCMV *peptide for three days. The cells were recovered and co-cultured with* BALB/c *derived macrophages in the presence of the peptide. During the last 12 hours of the co-culture,* 3H *thymidine was added to the culture. The cells were harvested and* 3H *thymidine incorporation was assessed. The highest incorporation was observed in* CD40 *deficient macrophage T cell co-culture.*

Reason: CD40 is an important molecule on B cells which is involved in cognate interaction between T and B cells. Macrophages actively phagocytose organism or even tumour cells invitro. Transduction of signals through CD40 induces upregulation of CD80/CD86 and thus helps to provide co stimulatory signals to responding T cells signalling through CD40 is also essential for germinal centre development and antibody response to T-cell dependent antigen.

In a stressfull condition, ACTH *secretion was increased and as a result glucocorticoid concentration was elevated in blood.*

Adrenocorticotropic hormone (ACTH)

ACTH is produced by pituitary gland. Its key function is to stimulate the production and release of cortisol from the cortex of the adrenal gland. Cortisol is a steroid hormone in the glucocortisoid class of hormones. During fasting, it stimulates gluconeogenesis (formation of glucose) and activates antistress and anti inflammatory pathways. It has an indirect role in liver glycogenolysis (breakdown of glycogen to glucose1- phosphate). Cortisol prevents the release of substance in the body that cause inflammation. Used to treat conditions that occur as a result of over activity of B-cell mediated antibody response. It inhibits production of IL-2, interferon (IFN)-gamma, Tumor necrosis factor (TNF)-alpha by antigen presenting cells (APC 1) and T helper (Th - cells) but upregulates- IL4, IL 10, IL 13 by Th2 cells. It can weaken the activity of immune system. It prevents proliferation of T cells by rendering IL 2 producer T cells, unresponsive to IL 2 and unable to produce IL 2 (T- cell growth factor). Cortisol has negative feedback effect on IL-1.

Chronically elevated levels of ACTH occur in primary adrenal insufficiency (Addison's disease)

Cushing's disease, a pituitary tumor is caused by elevated level of ACTH and excess of Cortisol.

Monoclonal antibodies (mAB) *can be potentially used as therapeutic agents. The major advantage is that they can specifically target aberrant cells. However, there is a practical difficulty. Monoclonal are raised in mouse and therefore it is expected that an immune reaction will develop if these are injected into humans. It is therefore necessary to humanize monoclonal antibody by taking a human* IgG *and replacing the* CDRs *by those derived from the mouse* mAB.

Theory: The process of humanization is usually applied to monoclonal antibodies developed for administration to humans. Humanization can be necessary when the process of developing a specific antibody involves generation in a non- human immune system (such as in mice). The protein sequences of a humanized antibody is essentially identical to that of the human variant, despite the non human origin of some of its complementarity determining region (CDR) segments responsible for the ability to bind to its target antigen. 'Direct' creation of a humanized antibody can be accomplished by injecting the appropriate CDR coding segments (responsible for the desired binding properties) into a human antibody 'scaffold'. This is achieved through recombinant DNA methods using an appropriate vector and expression in mammalian cells. That is, after an antibody is developed to have the desired properties in a mouse (or other non-human), the DNA coding for that antibody can be isolated, cloned into a vector and sequenced. The DNA sequence corresponding to the antibody, CDRs can then be determined. Once the precise sequence of the desired CDRs are known, a strategy can be devised for inserting these sequences appropriately into a construct containing the DNA for a human antibody variant. The strategy may also employ synthesis of linear DNA fragments based on the reading of CDR sequences. For humanization of mAB mouse constant region are replaced by their human counterpart

Ten different mouse strains were primed with whole keyhole limpet haemocyanin (KLH). *KLH was broken into ten peptides for invitro stimulation. The splenocytes from ten different primed moue strains were restimulated with each of these ten peptides and the responsiveness to these peptides were measured invitro. It was found that each of these mouse strains had responded to one of the peptides when the peptide 3 responder was mated with peptide 4 responder, the splenocytes of* F1 *offspring responded to both the peptides because mouse strain responding to peptide 3 or peptide 4 have different* MHC *haplotypes.*

Reason: Keyhole limpet Haemocyanin (KLH) is a protein that can be used to deliver vaccines to the body. Its large size and numerous epitopes generate a substantial immune response and abundance of lysine residues for coupling haptens, allows a high hapten: carrier protein ratio increasing the likelihood of generating hapten-specific antibodies. Using splenocytes from KLH-immunized mice showed that recombinant CTB (r CTB) did not affect the KLH specific proliferation of splenocytes isolated from mice immunized two weeks earlier. However rCTB strongly enhanced the KLH-specific proliferation of splenocytes from mice immunized with KLH 4 weeks prior. Using antibody co culture systems, it was shown that r CTB directly co stimulates KLH specific CD^4+ and CD^8+ T cell proliferation but not B cells enzyme linked immunospot (ELISPOT) assays revealed that rCTB also enhanced and KLH specific CD4+ T cell mediated production of interleukin 2(IL 2), IL 4and interferon gamma (IFN gamma) by four to five fold. Characterizing the adjuvant effect of rCTB invivo confirmed the results above viz., rCTB mediated a two fold increase in the ex vivo Tcell response when used as a classified adjuvant in a secondary, but not in a primary KLH immunization regimen. Together these data shown that rCTB can act as a strong adjuvant, by directly co stimulating antigen primed CD^4+ and CD^8+ T cell in a dose dependent manner.

Stem cells are widely used for their regenerative property and capacity to differentiate into different lineages. A person with a damaged liver approaches a stem cell therapist. The therapist suggested transforming skin cells from the patient into induced pluripotent stem cell (ips) cells and using them for further differentiation and grafting in liver.

Reason: Stem cells are defined as undifferentiated cells (lacking certain tissue specific differentiation markers) which are capable of proliferation and able to self maintain the population. Stem cells are of two types named embryonic stem cells and adult stem cells. Stem cell from one tissue may be able to give rise to cell types of a completely different tissue a phenomenon known as plasticity.

A mouse was primed with trinitrophenyl lipopolysaccharide (TNP-LP) *whereas another mouse was primed with* TNP *keyhole Limited hemocyanin*(TNP-KLH). *After three weeks, these nice were sacrificed and splenic cells were fractioned to* B *cells and* T *cells.* B *cells from*TNP-LPs *primed mice were cocultured with* T *cells from* TNP-LPs *or* TNP-KLH-*primed mice. Similarly* B *cells from* TNP-KLH *primed mice were co cultured with the* T *cells from* TNP-LPS *or* TNP-KLH *primed mice. The highest* IgG *production is found in co culture* $B^{TNP-KLH} \times T^{TNP-KLH}$

Reason: IgG is the major immunoglobin in normal human serum accounting for 70 to 75% of total immunoglobin pool. Trinitrophenylated lipopolysaccharide is thymus - independent immunogen. Spleen cells from mice pre injected with high doses of bacterial lipopolysaccharide did not generate anti trinitro phenyl (TNP) plague forming cells in vitro to the T - dependent antigen, TNP - sheep erythrocytes but did generate fully plague forming cells to the T-independent antigen, TNP-Ficoll and TNP - *Brucella abortus*.

Oncogenes and tumor suppressor genes are termed as cancer critical genes. Increasingly powerful tools are now available for systematically searching the DNA or mRNA of cancer cells for either significant mutation or altered expression. To identify the independently an oncogene or a tumor suppressor gene, the most convenient test is to use a transgenic mice that overexpress the candidate oncogene and knockout mice that lack candidate tumor suppressor gene.

Theory: There are two mutational routes towards the uncontrolled cell proliferation the first is due to oncogene and second due to tumor suppressor gene. A single oncogene in Mouse is not usually sufficient to turn a normal cell into cancer cell. This can be demonstrated by studies of transgenic mice.

Cystic fibrosis (CF) transmembrane conductance regulator protein (CFTR) is known to be a cAMP dependent chlorine channel. CF patients (with mutant CFTR proteins) show reduced chlorine permeability and as a result exhibit elevated chlorine level in sweat. To prove this, CFTR proteins (both wild type and mutant) are inserted in a model membrane (liposome) and chlorine transport is followed with radioactive chlorine. It is known that topology of CFTR in membrane is very important for its function. Despite no proteolytic degradation or denaturation of CFTR proteins, wild-type CFTR failed to transport chlorine in liposome because CFTR protein gets wrongly inserted in liposomes.

Reason: Cystic fibrosis disease is caused by a mutation in gene embodying an ABC transporter that functions as a chlorine channel in the plasma membrane of epithelial cells. The channel is unusual in that it requires both ATP hydrolysis and cAMP dependent phosphorylation.

In order to prevent tetanus in neonates one of the following treatment can be adopted

- Treatment of the Infant with antitoxin and the toxoid
- Immunize the mother with the toxoid

In the case of A the treatment can be given after the onset of the conditions and in the case of B the immunization has to be done late in the pregnancy.

Reason: The tetanus toxoid vaccine is given during pregnancy to prevent tetanus to mother. Antibodies formed in mother after vaccination passed to baby and protect the baby for a few month after but it also helps to prevent premature delivery.

Survival of intracellular pathogens depends on the levels of pro inflammatory and anti inflammatory cytokines in macrophages. In an experimental condition, Mycobacteria infected macrophages were treated with IL-6 or IL 12 for 4 hours at 37°C. Untreated cells were used as control. Cells were lysed and number of bacteria in each experimental set was counted by measuring colony forming unit (CFU). It was observed that IL 12 treated cells contain less intracellular bacteria than control.

Reason: In vitro and in vivo studies have demonstrated that cytokine interleukin12 (IL 12) plays an important role in the orchestration of the immune defence of the host against *Mycobacterium avium*. IL 12 is secreted by macrophages following infection and deficiency of cytokine is associated with accelerated progression of the disease.

You are given a group of four mice. Each mouse is immunized with keyhole limpet hemocyanin or azobenzene arsoante or lipo polysaccharide or dextran. Four weeks later, sera were collected from these mice and antigem specific IgG1 and IgG2a ELISA were performed. Only one of the mice showed positive response. It was keyhole limpet hemocyanin primed mouse.

Reason: Experimental auto immune Myasthenia gravis (EAMG) is an animal mode of human Myasthenia gravis (MG). In mice, EAMG is induced by immunization with torpedo California acetylcholine receptor (Ach R) in complete Freund's adjuvant (CFA). However, the role of cytokines in the pathogenesis of EAMG is not clear. Because EAMG is an antibody mediated disease, it is of the prevailing notion that Th2 but not Th1 cytokines play a role in the pathogenesis of this disease. To test the hypothesis that the Th1 cytokine, interferon (IFN)-gamma plays a role in the development of EAMG, it was immunized IFN- gamma knockout (IFN- gko)(- / -) mice and wild type (WT) (+ /+) mice of H-2(b) haplotype with AchR in CFA. It was observed that AchR- primed lymph node cells from IFN-gko mice proliferated normally to AchR and to its dominant pathogenic alpha 146-162 sequence when compared with these cells from the WT mice. However, the IFN –gko mice had no sign of the muscle weakness and remained resistant to clinical EAMG at a time when the WT mice exhibited severe muscle weakness and some died. The resistance of IFN gko mice was associated with gently reduced levels of circulating anti- AchR antibody levels compared with those in the WT mice. Comparatively, immune sera from IFN-gko mice showed a dramatic reduction in mouse AChR- specific IgG1 and IgG2 antibodies. However, keyhole limpet hemocyanin (KLH)- priming of IFN-gko mice readily elicited both T cell and antibody responses, suggesting that IFN- gamma regulates the humoral immune response distinctly to self (AchR) versus foreign (KLH) antigens. It is concluded that IFN-gamma is required for the generation of a pathogenic anti AchR humoral immune reponse and for conferring susceptibility of mice to clinical EAMG.

Tumor cells were isolated from a breast cancer patient. These cells were injected into nude mice and they were divided into four groups. Group 1 received EGF receptor conjugated with methotrexat. Group 2 received transferring receptor- conjugated with methotrexate; group 3 received mannose receptor conjugated with methotrexate; group 4 received same amount of the free drug. In the case of transferring receptor conjugated drug the tumorogenic index would be minimum.

Reason: Methotrexate affects cancer and rheumatid arthritis by two different pathways. It allosterically inhibits dehydrofolate reductase (DHFR) enzyme which participate in synthesis of tetrahydrofolate. Methotrexate inhibits synthesis of DNA, RNA, thymidylates and proteins. The transferrin receptor (TFf) are over expressed on tumour cells. The drug is selectively taken up by tumour cell via Tf-TfR interaction and leads to release of methotrexate in the cells.

CONCEPTS RELATED TO EVOLUTION
KEY EVENTS IN GEOLOGICAL TIME SCALE
Paleoproterozoic:
Sidesian period
- Iron band formation
- Formation of magnets
- Anaerobic algae produced oxygen as photosynthetic which resulted in the formation of oxygen rich atmosphere.

Rhyacian:
- Evolution of Eukaryotes (primitive multicellular organisms)
- Multicellular *Francevillian* group fossils are dated from this period.

Orosirian
- This era is marked by the two largest known event impacts. A large asteroid collision that resulted in the vredefort impact structure. This created the Sudbury basin structure Columbian.
- Second is assembling of the supercontinent Columbia termed Statherian

Proterozoic
This period is subdivided into Mesoproteozoic and Neoproteozoic eras

Mesoproterozoic
Calymmian:
- Expansion of existing platform covers
- Supercontinent Columbia broke up during the calymmian.

Ectasian:
- First evidence of sexual reproduction
- Hunting evolved on Somerset Island, Canada
- Microfossils of multicellular filaments of *Bangiomorpha pubescens* (red algae) which is the first taxonomically evolved eukaryote.

Stenian
- Supercontinent Rodinia was assembled on the Keweenawan rift.

Neoproterozoic

Tonian
- The supercontinent Rodinia was broken
- First large radiation of aeritarche
- First Metazoans (animals)- started to appear in the late Tonain.

Cryogenian
- Sturtian and Marinoan glaciations
- Sturtian is the greatest of the ice ages.
- These glaciations covered the entire planet (snowfall Earth) or a band of open sea survived near equator
- Superocean Mironia began to close while superocean Panthalasa began to form.
- Cratons later assembled into another supercontinent called Pannotia, in the Ediacaren
- Deposition of dolomite reduced Co_2.
- Testate amoeba fossils
- Oldest known fossil of sponges
- Red algae, green algae, Sramenopiles, Ciliates, Dinoflagellates and testate amoeba dated to this period.

Edicaran
- Named after Edicara hills of South Australia
- End of global Marinoan glaciations
- Carbonate layer
- Multicellular organisms
- Moon at this time meant that tides were stronger and more rapid.

Phanerozoic
This period is subdivided into three eras namely Paleozoic, Mesozoic and Coenozoic.

Paleozoic

Cambrian
- Change in life on Earth
- Prior to Cambrian majority of living organisms are small, unicellular and simple

- Complex, multicellular organisms called Metazoa (animals) and flagellated colonial Proiste similar to modern Choanoflagellate.
- Microbial biofilm
- Continents are dry and rocky due to lack of vegetation
- Shallow seas created during breakup of supercontinent pannotia
- Seas warm and polar ice is absent
- Laurentia (North America), Baltica and Siberiia are separated from Gondwana to form isolated land mass
- Soil formation

Ordovician

- Invertebrates Molluscs and Arthropods dominated the oceans
- Ordovician- Silurian extinction event
- Great Ordovician Bio diversification event.
- Fish, the world's first true vertebrates and those with jaws appeared.
- Fossil of the species belonged to this era mark the present years major petroleum and gas reservoirs
- Gondwana drifted toward South pole
- Rheic ocean formed between Gondwana and Avalonia
- Taconic orogony is the major mountain building episode.
- Ordovician meteor event but does not have any major extinction event.
- South poles are filled with ice caps
- Ice age at the end of Ordovician period due to the expansion of first terrestrial plants.
- Filter feeding organisms
- Articulate Brachiopods, Cepholopodis, Crinoids

Silurian

- Base for this is set at the major Ordovician- Silurian extinction events in which 60% of marine species are wiped out.
- Diversification of jawed and bony fish
- Bryophyte like vascular plants

- Terrestrial Arthropods
- Oldest known Tracheophytes of genus Cooksonia appear
- First terrestrial animals and are represented by air breathing Millipeds
- With the supercontinent covering the equator and much of the Sothern hemisphere, a large ocean occupied most of the northern part of the globe
- Number of island chains
- Melting of icecaps and glacier resulted in raise in sea level
- Avalonia, Baltica and Laurentia continents drifted together near the equator to form second supercontinent Euramerica
- Panthalasa is the vast ocean in the northern hemisphere
- Newly formed Ural ocean
- Earth entered long, warm green house phase supported by high CO_2 levels
- Moss like miniature forests along lakes and streams
- First fossil record of vascular plants with tissues that carry water and food appeared in the second half of the Silurian period.
- Earliest known representative of the above is Cooksonia.
- First bony fish Osteichthyes
- Animals adapted fully to terrestrial conditions appeared

Devonian

- First significant adaptive radiation of life on dry land occurred
- Free sporing vascular plants began to spread on the dry land
- Extensive forests formation which covered the continents
- Several plant groups evolved trees and true roots
- First seed bearing plants appeared
- Various terrestrial Arthropods established
- Age of fish
- Ray finned and lobe finned bony fish appeared.
- Primitive sharks more dominant in oceans
- Gondwana to the South, continent of Siberia to the north and the early formation of small continent Euramerica in between

- Greenhouse gas
- Warm period and lacked glaciers
- Co_2 level decreased
- The barrier reef, Kimberley basin of Northwest Australia of present are dated to this period.
- Reefs built by Corals and calcareous algae
- Rapid appearance of so many plant groups and growth forms has been called Devonian explosion.
- Arthropods solidly established on land
- Erosion and sedimentation resulted in soil formation
- Late Devonian extinction effect which affected marine community, reef barriers.
- Land plants and freshwater species remained unaffected.

Carboniferous (Mississippian / Pennsylvanian)

- Coal bearing
- Terrestrial life well established
- Amphibians -dominant vertebrate one branch of which would evolve into amniotes is the first solely terrestrial vertebrates.
- Arthropods are common and larger than current ones
- Forest cover which after death of plants laid down and converted to coal
- Atmospheric Oxygen reached highest of 35% which increased the size of terrestrial invertebrates to a great level (insects and arthropod)
- Glaciations, low sea level, mountain building as continents collide with Pangea
- Gondwana collided with Laurasia (north America and Europe)
- Hercynian orogeny in Europe and Alleghenian orogony in North America
- Eurasian plate welded to Europe along Ural mountains
- Apalachains extended Southward as Ouachita mountain
- Eurasian plate welded to Europe along Ural mountains
- Late Carboniferous Pangea shaped like 'O'

- Two oceans inside 'O' Pangea are Panthalasa and Paleo Tethys.
- Marine life is marked by *Cirnoids, Echinoderms, Brachiopods, Trilobites* (uncomon)
- The land is filled diverse plant population
- Vertebrates included large amphibians.
- Plants – *Equisetales, sphenophyllales, Lycopodiales* (Club mosses), *Lepidodendrales* (scale trees), *Filicales* (ferns), *Cordaitales, Cycads*, Seed ferns (Callistophytales), *Lepidodendron, Anabathra, Sigillaria*, the roots of these are known as *Stigmaria, Sphenophyllum* (slinder climbing plant), sponges (marine)

Permian
- Diversification of amniotes into ancestral form of mammals, turtles, *Lepidosaurus* and *Archosaurus*
- The two continents are Pangea and Siberia
- Ocean Panthalasa
- Desert within continental interior
- Ended with Permian-Triassic extinction event which resulted in largest mass extinction in Earth's history in which 90% of marine species and 70% of terrestrial species died out
- Dry condition favoured the growth of Gymnosperms.
- Ferns flourished in wetter environment
- First modern trees were Conifers, Ginkgos, Cycads appeared.
- Terrestrial life is dominated by diverse plants, fungi, Arthropods and many Tetrapods.
- Relatives of Cockroaches, ancestors of dragon flies and semi aquatic insects.

Mesozoic

Triassic
- Both start and end of this period marked by extinction.
- First true mammal ,Therapsida and first flying vertebrate, Pterosaurs are dated to this period.

- Climate was hot and dry and much of Pangaea's interior was marked by desert.
- End was marked by Triassic- Jurassic extinction event which wiped out many groups and allowed dinosaurs to assume dominance in Jurassic

Jurassic

- Two other extinction events:
- Pliensbachian/ Toascian event occurred in Early Jurassic
- Tithonian event at the end of the era
- Pangea drifted into Laurasia (North) and Gondwana (South)
- Climate ranged from dry to humid
- Dominated by dinosaurs
- First birds appeared and evolved from branch of Threopod dinosaurs
- Gulf of Mexico opened in new rift newrift between North America
- Calcite sea

Cretaceous

- Warm climate resulted in high eustatic sea levels
- Populated by now extinct marine reptile , ammonites and rudists
- Dinousaus dominated the land
- New group of mammals and birds as well as flowering plants appeared
- Ended with large mass extinction: Cretaceous-palcogene extinction event in which many groups including non avain dinosaurs and large marine reptiles died out

Coenozoic era

Paleocene

- Paleocene period is divided into three epochs namely Paleocene, Eocene, oliocene
- Small simple forms into large group of diverse animals called mammals

- Paleocene –Eocene period the thermal condition was maximum and resulted in significant global change that upset oceanic and atmospheric circulation and lead to the extinction of numerous deep sea benthic foraminifera and on land which is a major turnover in mammals.

Neogene
- This is divided into two epochs earlier Miocene and late Pliocene
- Mammals and birds evolved into modern form hominids- the ancestors of humans (Africa)
- North and South America are connected by Isthmus of Panama

Quaternary
- This is subdivided into two epochs named Pliestocene and Holocene
- This is the youngest period.

Heritability:

Heritability estimates the degree of variation in a phenotypic trait in a population viz., due to genetic variation between individuals in that population. Heritability measures the fraction of phenotype variability that can be attributed to genetic variation.

$H^2 = Var(G)/Var(P)$

$Var(G) \longrightarrow$ genetic variance and $Var(E) \longrightarrow$ Environmental variance

A particularly important component of the genetic variance is the additive variance Var (A), which is the variance due to average effects (additive effects) of the alleles. Since each parent passes a single allele per locus to each offspring, parent-offspring resemblance depends upon the average effect of single alleles. Var (A) represents the genetic component of variance responsible for parent offspring resemblance. The additive genetic portion of the phenotypic variance is known as narrow sense heritability and is defined as

$h^2 = Var(A)/Var(P)$

upper case H^2 is used to denote broad sense and lower case h^2 for narrow sense.

Var(A) is important for selection

Comparison of relatives:

$h^2 = b/r = t/r$

r – coefficient of relatedness

b – coefficient of regression

t – coefficient of correlation.

Calculation of realized heritability as the response to selection relative to the strength of selection

$h^2 = R/S$

De Friesfulher method is used for analyzing twins.

Twin studies

Fraternal or dizygotic (DZ) twins on average share half their genes and so identical or monozygotic (MZ) twins on average are twice as genetically similar as DZ twins. A crude estimate of heritability then is approximately twice the difference in correlation between MZ and DZ twins

Falconer's formula,

$H^2 = 2\,[r\,(MZ) - r\,(DZ)\,]$

The effect of shared environment C^2 contribute to similarity between siblings due to the commonality of the environment they are raised in. Shared environment is approximately by the DZ correlation minus half heritability, which is the degree to which DZ twins share the same genes

$CZ = DZ - h^2/2$

Unique environmental variance e^2, reflects the degree to which identical twins are raised together and dissimilar

$e^2 = 1 - r(MZ)$

Selective breeding of plants and animals the expected response to selection of a trait with known narrow sense heritability (h^2) can be estimated using the breeder's equation

$R = h^2 S$

R – Response to selection

S – Selection differential

Cladogram

Cladogram depicts a sequence in the origin of derived characters. It is interpreted as a family tree depicting a hypothesis regarding monophyletic lineage. The following terms are used to identify shared and distinct characters among groups.

Pleiomorphic characters

These are present at the base of a tree viz., ancestral state.

Apomorphic characters

Apomorphic characters (separate form) are characters believed to have evolved within the tree. It can be used to separate one group in the tree from the rest.

Homoplasy

Characters shared by members of a tree but not present in their common ancestor.

Fecundity

Fecundity is the reproductive output, usually of an individual or number of offspring produced or capacity of reproduction.

Hymenopteran insects

In Hymenopteran insects, male are haploids and females are diploid. All fertilized eggs give rise to females and all unfertilized eggs give rise to males. As a result, if female mates with a single female, the female in the progeny are related to each other by 75% but if the female mated with many males the mean genetic relatedness of female progeny decreases. If a female mates with a single male, the female in the progeny are more genetically related as compared to the female progenies by mating of one mother with many males. In short, the percentage of relatedness decreases with increase in the number of males.

Time periods of Marsupials

Late Jurassic: Early Therians arrived in Antarctica- Australia where the Marsupials subsequently evolved.

Paleocene: Marsupials entered Australia from South East Asia

Eocene: Chance dispersal of Marsupials into Australia.

Marsupials are mammals in which the young are borne early in their development. Marsupials and placental mammals were separated from each other about 120 mya. Marsupials were initially widely distributed throughout Gondwana, which split up about 100 mya. The island continent of Australia because of its early isolation by sea escaped from placentals. The Australian Marsupials evolved into a wide variety of forms.

Types of natural selection

Stabilizing selection

When selective pressure is to select against the two extremes of a trait, the population experiences stabilizing selection. For example, when a plant that is too short may not be able to compete with other plants for sunlight. However, extremely tall plants more susceptible to end damage.

Directional selection

In directional selection, one extreme of the trait distribution experiences selection against it. The result is that the population trait distribution shift toward the extreme. In case of such selection, the mean population graph shifts. Example: The neck of a giraffe

Disruptive selection

In this, selection pressure acts against individuals in the middle of the trait distribution. The result is a bimodal or two peaked, curve in which the two extremes of the curve create their own smaller curves. For example, image a plant of extremely variable height that is pollinated by three different pollinators, one that was attracted to short plants, another that preferred plants of medium height and the third that visited only the tall plants. If the pollinator that preferred plants of medium height disappeared from an area, medium height plants would be selected against and the population would tend toward both short and tall but not medium height plants. Such a population in which multiple distinct forms or morphs exist is said to be polymorphic.

Panmictic population

The approximate effective population size in a panmictic population of 240 with 200 males and 40 polygamous males is

$Ne = 4 N_f N_m / N_f + N_m$

$= 32000/240$

$= 133$

Theory: An isolated group of individuals living together in an area and freely interbreed among themselves has been termed as Panmictic population. The effective size of a population, Ne determines the rate of change in the composition of a population caused by genetic drift, which is the random sampling of genetic variants in finite population. Ne is crucial in determining the level of variability in a population, and the effectiveness of selection relative to drift. In particular, the action of selection means that Ne varies across the genome and advances in genomic techniques are giving new insights into how selection shapes Ne. For species in which males have a harem (Example: the northern hemisphere elephant seal), Ne is strongly affected by sex-ratio bias.

Why is it difficult for selection to eliminate a completely recessive deleterious allele from a population?

It is difficult for natural selection to eliminate a completely recessive deleterious allele from a population because at low frequencies most of the alleles will be in heterozygous state where they will have no phenotypic effect. As the allele gets increasingly rare the frequencies of homozygous genotypes becomes even lower giving selection little opportunity for selection to act against the deleterious allele. Similarly, a new recessive beneficial mutation is initially found in the heterozygous state where the beneficial phenotype is not expressed. Only when the homozygous recessive genotype is formed selection can increase the frequency of this allele. Initially this takes sometime but once the frequency increases a bit it takes off.

Imprinting

Imprinting is a learning process that occurs at a remarkably early age. Many animals pass through time periods in their individual growth when certain learning experiences are far more significant than at other times. Such as specific period is referred to as critical period.

Example 1: The number of trials required for rats to learn a task when they were exposed to various conditions was as follows:

Experimental condition	Observation
Light:Light darkcycle – 12h:12h	N-trials
Bright light -24h	Significantly more trials than N
Bright light – 24 h+ continuous physical disturbance	Significantly more trials than N
Dark light – 24 h+ continuous physical disturbance	Significantly more trials than N

Inference: Learning was reduced by sleep loss

Example 2: Assume a male sparrow (species X) is hatched and reared in isolation and allowed a critical imprinting period to hear the song of a male of another sparrow (Species Y). Now after the isolation it will sing the song of species Y that it had heard in the critical period.

Aggressive behavioural increases are expected in a rat when its nucleus accumbens is experimentally ablated.

Reason: The nucleus accumbens is a collection of neurons and forms the main part of the ventral striatum. It is thought to play an important role in reward, pleasure, addition, aggression, fear etc.,

Suppose you discovered a new species about which you know only two facts: it is small sized (<10 cm) and short lived (< 20 days) we could predict that the individuals of that species breeds early and only once in life and produces large number of small sized offspring.

Reason: Selection acts on phenotype, the physical, biochemical and behavioural traits of organisms. In unstable r selection predominates as the ability to reproduce quickly. Traits that are thought to be characteristic of r selection include high fecundity, small body size, early maturity, short generation time and ability to disperse offspring widely.

One hundred independent populations of Drosophila are established with 10 individuals in each population, of which one individual is of Aa genotype and the other nine are of AA genotype. If random genetic drift is the only mechanism acting on these populations, then after a large number of generations, the expected number of populations fixed for the 'a' allele is 50%

Reason: Assuming genetic drift is the only evolutionary mechanism acting on an allele at any given time the probability that an allele will eventually present in the population is simply its frequency in the population at that time. As in this question gene frequency is not given, assume that, $p = q = 0.5$. As the gene frequency of $a = 0.5$, then the probability that 'a' will become fixed is 50% (0.5 1000= 50)

Characteristics that make Amborella the most basal living Angiosperm are the presence of carpels and absence of vessel elements.

Amborella's lineage diverged from other Angiosperms around 130 million years ago, sometimes after the first flowering plant appeared. *Amborella* has all of the defining features of a flowering plant, but at the same time it seems to have retained some 'gymnospermy' characteristics as well. For instance, it lacks the vessel elements for water conduction present in most other flowering plants. Also while

Amborella has carpels, they are incompletely closed. This is significant because the carpel is thought to have originated from a flat, leaf like structure with ovules on its margins. This structure eventually rolled toward and become enfolded, creating a hollow, enclosed ovary with one or more ovules. Early angiosperms probably had carpels that were not quite fused shut but were sealed with secretions from the carpel, which is the case with *Amborella*.

In life history evolution there is generally a trade off between the sizes and number of offspring produced. The conditions that would favour the production of a small number of large sized offspring are provision of parental care and predator's preference for large sized prey.

Life History theory is an analytical framework designed to study the diversity of life history strategies used by different organisms throughout the world. A life history strategy is the age and stage specific patterns and timing of events that make up the organism's life such as birth, weaning, maturation, death etc.,

List of traits

- Size at birth
- Growth pattern
- Age and size at maturity
- Number, size and sex ration of offspring
- Age and size specific reproduction investments
- Age and size specific mortality schedules
- Length of life

Reproductive Value: Models those tradeoffs between reproduction, growth and survivorship.

RV= current reproduction+ Residual reproductive value (organisms further reproduction through its investment in growth and survivorship)

Cost of reproduction hypothesis: Higher investment in current reproduction hinders growth and survivorship and reduces future reproduction while investment in growth will pay off with higher fecundity and reproduction episodes in the future.

r/k selection theory: Central trade off to high history is the number of offspring versus timing of reproduction. Organisms that are r selected have a high growth rate (r) and tend to produce a high number of offspring with minimal parental care; their life spans are tending to be shorter. r selected organisms are suited to life in an unstable environment, because they reproduce early and abundantly and allow for a low survival rate of offspring. k-selected organisms subsist near the carrying capacity of their own environment (k), produce a relatively low number of offspring over a longer span of time and have high parental investment. They are more suited to stable environment in which they can rely on a long lifespan and a low mortality rate that will allow them to reproduce multiple times with high offspring survivor rate. Some organisms that are very r-selected are semelparous, only reproducing once before they die. These may be short lived like annual crops. Some semelparous organisms are relatively long lived such as African flowering plant *Lobelia telekii* which spends upto several decades growing an inflorescence that blooms only once before the plant dies or periodical cicada which spends 17 years as a larva before emerging as an adult. Organism with longer life spans are iteroparous (are more r-selected than k-selected example: sparrow), reproducing more than once in a lifetime.

r-selected organisms

- mature rapidly and have an early age of first reproduction
- Have a relatively short life span
- Have a large number of offspring at a time and few reproductive events, or are semelparous having high mortality rate and a low offspring survival rate
- Have minimal parental care/investment

k-selected organisms

- Mature more slowly and have a later age of first reproduction
- Have a longer life span
- Have few offspring at a time and more reproductive event spread over a longer span of time and have a low mortality rate and a high offspring survival rate
- Have high parental investment

Assume that individual A wants to do an altruistic act to individual B and that the benefit and cost of doing this act are in 'fitness' units 40 and 12 respectively. According to Hamilton's rule, A should perform the altruistic act only if B is his grandson or grand daughter

Benefit = 40, cost = 12

C/B< r

12/40. < r

=0.3 which is approximately near o.25

Hence concluded as grandson or grand daughter

Theory: Sometimes, animals engage in apparent altruism (they exhibit behaviour that increases the fitness of other individuals by engaging in activities that decreases their own reproductive success). The key insight to understand the evolution of such self sacrificed behaviour was provided by British evolutionary biologists Hamilton. He argued that natural selection favours genetic success, not reproductive success per selection and such individuals can pass copies of their genes onto further generation. Genes are passed from direct parentage (the rearing of offspring and grand offspring) and by assisting the reproduction of close relatives (such as nephew, niece), a concept referred to as 'inclusive fitness' or 'kin selection'. Hamilton devised a formula known as Hamilton's rule that specifies the conditions under which reproductive altruism evolves.

r ×B>C

where B is the benefit (in number of offspring equivalents) gained by the recipient of the altruism. C is the cost (in number of offspring equivalents) suffered by the donor while undertaking the altruistic behaviour, and is the genetic relatedness of the altruist to the beneficiary.

Example: Female lion well nourished cub gain inclusive fitness by nursing a starting cub of a full sister because the benefit to her sister (B= one offspring that would otherwise die) more than compensates for the loss to herself (C= approximately one quarter of an offspring), since the survival probability of her own, non starving cub is only slightly reduced. Given that the average relatedness (r) between two full sister is 0.5, then according to Hamilton's rule $(0.5 \times 1) > 0.25$

Parent and siblings =r = 0.5

Grandparents, niece and nephew = r= 0.25

In diploid organisms every parent transmits 50% of its genetic information to each offspring. Siblings therefore share half of each Parent's contribution to their genome, adding to the coefficient of relatedness r = 0.5

Cousins share r = 0.125 or r = 1/8

Cousins are related to their common grandparents by ¼ = 0.25

Therefore an organism shares 50% of their genetic information with their parents, 50% with their siblings and 25% with uncle, aunt, grandparent and grand children.

Northern elephant seals had been reduced to about 20 individuals in the 1800s. Biologists studied variation in the proteins in the species. They found no genetic differences in the protein among different individuals. This lack of variation is due to population bottle neck and genetic drift

Reason: In small population, the frequencies of particular alleles may change drastically by chance alone such change in allele frequencies occur randomly as if the frequencies were drifting and are thus known as genetic drift.

Even if the organisms do not move from place to place yet, occasionally their populations may be drastically reduced in size. This may occur due to flooding, drought, and epidemic disease and may also be due to natural forces or from progressive changes in the environment. The few surviving individuals may constitute are random genetic samples of the original population. This resultant alternations and loss of genetic variability has been termed as bottle neck effect.

CONCEPTS RELATED TO EMBRYOLOGY

Cleavage in egg types:

Egg types	Cleavage
Telolecithal	Meroblastic
Isolecithal	Holoblastic
Centrolecithal	Superficial meroblastic
Mesolecithal	Holoblastic equal

Plant development

As plant development proceeds, cells with multiple potentials are mainly restricted to meristem regions. During embryogenesis the root-shoot axis develops, cell differentiation occurs and three basic tissue system or established. Three basic tissues differentiate while the plant embryo is in the globular stage. The root and shoot both arises from the apical meristem but their formation is independently controlled. Morphogenesis results from changes in planes and rates of cell division. Lateral roots develop from pericycle. In many vascular plants, secondary growth is the result of the activity of the two lateral meristems, the cork cambium and vascular cambium. Arising from lateral meristems, secondary growth increases the girth of the plant root or stem rather than its length. As long as the lateral meristems continue to produce new cells, the stem or root will continue to grow in diameter. In woody plants, this process produces wood.

Amniotic eggs

The earliest members of the class Reptilia were the first vertebrates to possess amniotic eggs. The amniotic eggs have extra embryonic membranes that protect the embryo from desiccation, cushion the embryo, promote gas transfer and store waste material. Amniotic eggs are found in reptiles, birds and mammals. Birds and mammals are called endotherms because they obtain their body heat from the cellular processes.

During early cleavage of Caenorabditis elegans embryos, each asymmetrical division produces one founder cell which produces differentiated descendants and one stem cell. The very first cell division produces one anterior founder cell namely AB and one posterior stem cell namely P1. When these blastomeres are experimentally separated and allowed to proceed further with development P1 cells would develop autonomously while AB would show conditional development.

Reason: *Caenorabolitis elegans* is a free living solid nematode having a rapid period of embryogenesis (16 hours) and reproduces by self fertilization or cross fertilization due to infrequently occurring males. The zygote exhibit rotational holoblastic cleavage. During early cleavage each asymmetrical division produce one founder cell (denoted AB, MS, E, C & D) which produces differential descendants and one stem cell (P1-P4 lineage). In the first cell division, the cleavage furrow is located asymmetrically along the anterior-posterior axis of the egg, closer to what will be the posterior pole. It forms a founder cell (AB) and stem cell (P1). During second division, the anterior founder cell (AB) divide equatorially (longitudinally, 90° to the anterior posterior axis) while the P1 cell divide meridonally (transversely) to produce another founder cell (EMS) and a posterior stem cell P2. The stem cell lineage always undergoes meridonial division to produce anterior founder cell and posterior stem cell. The posterior stem cell will continue the stem cell lineage.

Conditional and autonomous division: When stem cell differentiate, their fate may be determined autonomously (no external cues are required) or conditionally (when their fate is determined by external factors). Hence, with the first division of the zygote, there are two cells, the P1 cell and the AB cell. The P1 cell is able to make all of its fated cells while the AB cell can only make a portion of the cells it was fated to produce. Thus the first division gives the autonomous specification of two cells, but the AB cell requires a conditional mechanism to produce all of its fated cells.

In the case of sea urchin, the correct sequence of events taking place during the interaction of sperm and egg are,

- *Chemo attraction of sperm to the egg by soluble molecules secreted by the egg*
- *exocytosis of the sperm's acrosomal vesicle to release its enzymes*
- *binding of the sperm to the extracellular matrix of the egg*
- *passage of sperm through this extracellular matrix*
- *fusion of egg and sperm cell membrane*

Acrosome reaction

During fertilization, a sperm must fuse with the plasma membrane and then penetrate the female egg in order to fertilize it. Fusing to the egg usually cause little problem, whereas penetrating through the egg's hard shell or extracellular matrix can present more of a problem to the sperm. Therefore the sperm cells go through a process known as acrosome reaction which is the reaction that occurs in the acrosome of the sperm as it approaches the egg. The acrosome is a cap like structure over the anterior half of the sperm's head.

As the sperm approaches the zona pellucid of the egg, which is necessary for initiating the acrosome reaction, the membrane surrounding the acrosome fuses with the plasma membrane of the sperm's head exposing the content of the acrosome. The contents include surface antigens necessary for binding to the egg's cell membrane and numerous enzymes which are responsible for breaking through the egg's tough coating and allow fertilization to occur.

Variations in Acrosome reaction

Echinoderms

Some lower animal species a protuberance (acrosomal process) forms at the apex of the sperm's head, supported by a core of actin microfilaments. The membrane at the tip of the acrosomal process fuses with the egg's plasma membrane. In some Echinoderms including star fish and Sea urchin, a major portion of the exposed acrosomal content contains a protein that temporarily holds the sperm on the egg's surface.

Mammals

The acrosome reaction releases hyaluronidase and acrosin, their role in fertilization is not clear. The acrosomal reaction does not begin until the sperm comes in contact with the oocyte's zona pellucida. Upon coming into contact with zona pellucida, the acrosomal enzymes begin to dissolve and the actin filament comes into contact with zona pellucida. Once the two meet, a calcium influx occurs, causing a signalling cascade. The cortical granules inside the oocyte then fuse to the outer membrane and a transient fast block reaction occurs.

Steps in acrosomal reaction:

- The acrosomal reaction normally takes place in the ampulla of the fallopian tube when the sperm penetrate the secondary oocyte
- A few events precede the actual acrosome reaction
- Sperm cell acquire hyperactive motility pattern, by which its flagellum produces vigorous whip like movements, that propel the sperm through the cervical canal and uterine cavity, until it reaches the isthmus of fallopian tube.
- Sperm approaches the ovum with the help of various mechanisms including chemotaxis.

- Glycoproteins on outer surface of sperm bind with glycoprotein of zona pellucida of ovum.
- First stage is the penetration of corona radiate by releasing hyaluronidase from acrosome to digest cumulus cells surrounding the oocyte. The cumulus cells are embedded in a gel like substance made primarily of hyarulonic acid.
- After reaching Zona pellucida, the actual acrosome reaction begins.
- Acrosin digest the zona pellucida and membrane of the oocyte. Part of the Sperm's cell membrane then fuse with the egg cell's membrane and the contents of the head sink into the egg (lock and key mechanism).
- Zona pellucida also releases Calcium granules to prevent additional sperm from binding
- This binding is what triggers the acrosome to release the enzyme that allow the sperm to fuse with the egg
- As the sperm arrive at the egg, it is the egg that chooses the sperm and pulls it toward her.
- The selected sperm actually tries to swim away from the egg but is tethered to the egg by female hormones. The membrane around the egg literally opens up and swallows the sperm.
- Upon penetration, the process of egg activation occurs and the oocyte is said to have become activated.

Injection of noggin mRNA into a 1- cell UV - irradiated embryos of frog completely rescues dorsal development and allows the formation of complete embryo, this is because Noggin is a secreted protein which induces dorsal ectoderm to form neural tissue and it dorsalizes the mesoderm cells which would otherwise contribute to ventral mesoderm.

Noggin(NOG)

Noggin is a secreted protein responsible for the development of nerve tissue, muscle and bones. The molecular functions performed by NOG protein include protein homodimerization activity, protein binding, cytokine binding.

Noggin is a signalling molecule which promotes somite patterning. Released by notochord and regulate BMP during development. Absence of BMP4 will cause the patterning of the neural tube and somites from neural plate in developing embryo. It also cause formation of head and dorsal structures. Secreted noggin encoded by NOG gene binds and inactivates the transforming growth factor beta (TGF-β) super family signalling proteins like BMP4.

Formation of neural tissue, the notochord, hair follicles, eye structures arise from the ectoderm germ layer. Noggin activity in the mesoderm gives way to the formation of cartilage, bone and muscle growth and in the ectoderm noggin is involved in lung development.

> Somites are the blocks of mesoderms that are located on either side of the neural tube in developing vertebrate embryo.

Biological processes associated with Noggin

- Positive regulation of glomerules development
- Pattern specification process
- Axial mesoderm development
- Skeletal system development
- Cell differentiation
- Uteric bud development
- Negative regulation of astrocyte differentiation
- Pituitary gland development
- Positive regulation of branching involved in uteric bud morphogenesis
- Lung morphogenesis
- Negative regulation of cartilage development
- Somatic stem cell population maintenance

- Fibroblast growth factor receptor signalling pathway involved in neural plate anterior or posterior pattern formation.
- Positive regulation of epithelial cell proliferation
- Negative regulation of cytokine activity
- Neural plate morphogenesis
- Negative regulation of apoptotic signalling pathway
- Nervous system development
- Negative regulation of cardiac muscle cell proliferation
- Noggin is able to bind to BMP2, BMP4 and BMP7

In the context of the proximal distal growth and differentiation of a tetrapod limb some experiments were visualized that would show the possible interactions between the AER and the limb bud mesenchyme directly beneath it during limb development. They are

- *If the apical ectodermal ridge (AER) is removed at any time during the limb development, further development of distal limb skeletal element caeses.*
- *If leg mesenchyme is placed directly beneath the wing AER, distal hindlimb structure develop at the end of the limb.*
- *If an extra AER is grafted onto an existing limb bud, super numeracy structures are formed usually at the distal end of the limb.*

Theory: The paired forelimb and hindlimb of tetrapod are derived from the epidermal ectoderm and the mesoderm of somites and lateral plate. The development of limb is started by thickening of lateral plate mesoderm just beneath the epidermis of a presumptive limb area which may lie either behind the branchial region (for forelimb) or just in front of anus (for hindlimb) soon the epidermis lying over the mesenchyme mass becomes slightly thickened and bulged outward to form limb bud.

The Fate of a cell or a tissue is specified when it is capable of differentiating autonomously and being placed in a neutral environment with respect to the development pathway. An embryo will show a development patent based on its type of specification, based on this fact it could be said that the potency of a cell is greater than its normal fate in regulative development and is equal to its normal fate on mosaic development.

Reason: The blastula processes polarity and bilateral symmetry. It contains cell areas which in normal development become the germ layer and given rise to different parts of the embryo. Mosaic theory of development holds that the ontogenetic development consist of differentiation of cells from a more or less undifferentiated and homogenous egg with crude and simple pre localization of materials to greater differentiation and specialization of cells as development went on.

In an experiment, the sperm removed from the epididyms of a male mouse was added in a dish containing appropriate media and oocyte. No fertilization was seen. However when the sperm from epididymis was directly placed in uterus of an ovulated female, she became pregnant this observation suggested that, the contents of female reproductive tract interact with sperm and activate it for fertilization.

Reason: Fertilization involves fusion of male and female gametes. A chemical substance fertilizin secreted from the cortical region of the egg cytoplasm of sea urchin egg was supposed to attract the sperm. The movement of spermatozoa from the site of deposition to the site of fertilization usually depends upon active swimming of spermatozoa in body fluid by muscular contraction of female track.

The functionality of the pax 6 gene in the formation of optic and nasal structures may be attributed as:

Pax 6 renders the head ectoderm competent to receive signals from the optic vesicle. Apart from the optic vesicle, the head ectoderm may also be induced by BMP4 and FGF 8, so pax 6 is not exclusively for lens formation.

Theory: Pax 6 is also known as an iridia type II, protein or ocular hombin is a protein in human encoded by pax 6 gene involved in development of eyes. Certain neural and epidermal tissues which are ectodermal in origin

A two celled embryo is made of blastomeres A and B. If the two blastomeres are experimentally separated, the 'A' blastomere generates all the cells it would normally make. However the 'B' blastomere in isolation makes only a small fraction of cells it would normally make.

It could be concluded that 'A' blastomere is autonomously specified while 'B' blastomere is conditionally specified.

Sex determination in Drosophila

Sex determination in *Drosophila* is controlled by a cascade of genes whose expression is regulated by alternative splicing. Master switch gene in the hierarchy is sex lethal. Sex lethal is turned on only in females and an auto regulatory feedback loop which controls alternative splicing maintains this state. Sex lethal also promotes female differentiation by controlling the splicing of RNA from the next gene hierarchy transformer.

A mutant embryo of Drosophila in which one of the major sex determining gene, sex lethal, can only undergo default splicing, was allowed to develop.

The possible results of the above mentioned experiment are:

- ➢ The embryo will develop into a female fly

> Sex lethal gene product directly regulates sex specific alternate splicing of double sex RNA

During fertilization in mammals, Sperm-egg interaction is mediated by zona pellucida (ZP) *membrane proteins and their receptors present in sperm membrane.* ZP3 *has been identified to be the principle* 2P *protein whose post-translational modification is importance for sperm-egg interaction. In a competitive inhibition assay the sperm is saturated with either active* ZP3 *or its modified forms, before studying sperm egg interaction. Deglycosylating the* ZP3 *and using it for saturation of the sperm will not inhibit sperm egg interaction.*

Reason: This is because Zona pellucid sperm binding protein 3(ZP 3) or Zona pellucid glycoprotein 3 or the sperm receptor which binds sperm at the beginning of fertilization and is composed of 3 to 4 glycoproteins. Saturating the ZP3 protein prior to use, phosphorylating or dephosphorylating the ZP3 protein prior to saturation will inhibit the sperm egg interaction.

In the case of morphallatic regeneration there is re patterning of the existing tissue with little new growth.

Theory: In a morphallactic regeneration the new individual is produced not by addition of parts to the residue of animals body but by remodelling the entire available mass of cells into a new whole. This type of repair involve reorganization of remaining parts of the body of animal. For example, the regeneration of new individual from body pieces of Hydra, Planaria and earthworm.

With respect to the extra embryonic structures formed in the mammals, the possible functional attributes have been designated; amnion is a water sac and protects the embryo and its surrounding amniotic fluid. The epithelium is derived from Somatopleure. Chorion is essential for gas exchange in amniotic embryos. It is generated from the splanchnopleure.

Theory: The growing embryo develops four membrane called extra embryonic or foetal membranes. These are

Chorion: It surrounds the embryo, protect it and takes place in the formation of placenta

Amnion: The amniotic fluid present between amnion and embryo prevents desiccation of embryo and acts as protective cushion that absorbs shock.

Allantois: In human it is small and non functional except for furnishing blood vessels to the placenta.

Yolk sac: it is believed that it function as site of early bound cell formation.

CONCEPTS RELATED TO SYSTEMATICS AND ECOLOGY

Monocots	*Dicots*
Embryo with single cotyledon	Embryo with two cotyledons
Pollen with single furrow or pore.	Pollen with three furrows or pores
Flower parts in multiples of three	Flower parts in multiples of four or five
Major leaf veins parallel	Major leaf veins reticulated
Stem vascular bundles scattered	Stem vascular bundles in a ring
Roots are adventitious	Roots develop from radical
Secondary growth in absent	Secondary growth is present.

Type Specimen

Holotype: When a single specimen is clearly designated in the original description

Paratype: Additional specimen to the original specimen

Allotype: Specimen of the opposite sex to the holotype

Neotype: Specimen later selected to serve as the single type specimen when an original holotype has been lost or destroyed or where original author never cited a specimen.

Lectotype: specimen later selected from the original material to serve as the nomenclatural type when the holotype was not designated at the time of publication.

MacArthur and Wilson's equilibrium theory

MacArthur and Wilson coined the term island biography in their theory. Island biography examines the factors that affect the species of isolated natural communities.

Predictions of the theory:

- Islands close to a source area should have higher number of species than islands farther from the source area for islands of equivalent areas.
- Larger islands should have more species than smaller islands.
- Islands that are more isolated are less likely to receive immigrants than islands that are less isolated.

Primary production in aquatic ecosystem is measured using Light and dark bottle technique. In this method, an indirect measure of photosynthetic production dissolved oxygen concentration of the pond water enclosed in a BOD bottle is measured initially (I) and after a fixed duration of incubation in a light bottle (L) and a dark bottle (D). Then the gross and net primary production is estimated as

(L-D) *and* (L-I) *respectively.*

Gross primary productivity is the total rate of photosynthesis including the organic matter used up in respiration during the measurement period. While the net primary productivity is the rate of storage of organic matter in plant tissue in excess of respiratory utilization by plants during measurement period.

Respiratory activity = IB – DB

Net photosynthesis = LB-IB

Gross photosynthesis = Net primary production + Respiration

$$= (LB - IB) + (IB-DB)$$

$$= LB-DB$$

LB- Light bottle

DB- Dark bottle

IB- Initial bottle

Gause's competitive exclusion principle states that two species with identical niches cannot coexist indefinitely. According to this, the most appropriate statement regarding the validity of the principle is that it depends on how one defines niche.

Theory: Ecological Niche of an organism includes the physical space is occupied by its functional role in the community and conditions of existence. The validity of the Gause's competitive exclusion principle depends on the niche definition. According to Gause's competitive exclusion principle, one species have slightest advantage over another and the one with advantage will dominate. This leads to either exclusion of the competition or to an evolutionary or behavioural shift towards different ecological niche. In short, complete competitors cannot co exit.

Lotka - Voltera equations or Predator - prey equation

Population change through time according to the pair of equations

Prey $\quad dx/dt = \alpha x - \beta xy$

Predator $\quad dy/dt = \delta xy - \gamma y$

Where x denotes the number of prey; y denotes the number of predators; t denotes the time and $\alpha, \beta, \delta, \gamma$ denotes the positive real parameters describing interactions of two species.

- Prey population finds ample food at all times
- Food supply of the predator population depends entirely on the size of prey population
- The rate of change of population is proportional to its size
- During the process, the environment does not change in favour of one species and genetic adaptation is inconsequential
- Predators have limitless appetite.

Bottom up and Top down effects

In a lake ecosystem, bottom up effects (B) refer to control of a lower tropic level by the higher tropic levels and top down effects (T) refer to the opposite. The transfer of food energy from the producers through a series of organisms with repeated eating and being eaten is termed as food chain. In a lake ecosystem, the phytoplankton and zooplankton are controlled by bottom up effect and carnivore by top down effect.

Zeitgeber and entrainment

An animal was first maintained in a constant environmental condition for several days until a consistent biological rhythm (B) was established. The animal is then exposed to an experimental physical rhythm (E) which modulate the phase and period of B. However upon withdrawal of E, the B gradually regained its pattern of pre-exposure condition. This suggest that E is a zeitgeber and E cause entrainment of E.

Theory: A Zeitgeber is any external or environmental cue that entrains or synchronizes organisms biological rhythms to the Earth's 24 hour light or dark cycle and 12 month cycle. The setting up of the biological rhythms by environmental cues is called entrainment.

In the following statements taken from a research paper what does P in the parenthesis stands for the mean temperature of this region now is significantly higher than the one 50 years ago ($P<0.05$, t test).

The decision to accept or reject null hypothesis is based on sample study. There are two possible types of error in the test of hypothesis in research paper

Type 1- rejection of null hypothesis which is true

Type 2- acceptance of false null hypothesis

The P in the research paper stands for probability of T test difference in mean annual temperature of the two time periods

Autotrophs in the aquatic ecosystem, unlike their counterparts in the terrestrial ecosystem are mostly microscopic and very low in indigestible (to the herbivores) matter. This explains the fact that compared to the terrestrial ecosystem, in the aquatic ecosystem productivity / biomass ratios are higher and energy transfer rates tom higher tropic levels are faster.

Reason: The water covers about three fourth of the earth. Most of the aquatic body is occupied by algae, viz. The productivity / biomass ratio is higher in aquatic ecosystem.

Ecological compression differs from character displacement in that it operates on a shorter timescale and does not involve heritable change.

Theory: Character displacement refers to the phenomenon where differences among similar species whose distribution overlap geographically are accentuated in regions where the species co occur but are minimized or lost where the species distribution overlap. This pattern results from evolutionary change driven by competition among species for limited resources (example: food). The rationals for character displacement stems from the competitive exclusion principle also called Gause's law, which contends that to coexist in a stable environment two competing species must differ in their respective ecological niche; without differentiation, one species will eliminate or exclude the other through competition. Ecological compression means that species should specialize in terms of habitat they occupy.

Types of species in a community

- The species that has a large effect on community because of its abundance is termed as dominant species.
- The species that has a large role in community out of proportion to its abundance is termed as keystone species.

- Status of species that provide information on the overall health of the ecosystem is termed as indicator species.
- Significant conservation resource are allocated to a species which is single, large and instantly recognizable is termed as flagship species

Himalayan glaciations theory

Himalayan glaciations theory attributed distribution of species during the ice age. The former extension of the Himalayan glacier has been shown to have been considerable and the occurrence of Himalayan plants and animals on the higher ranges of Southern India may be due to the retreat of these species in the place towards the equator and as the temperature increased to the higher parts of the hills.

Age structure of a population

In most populations individuals are of different ages. The proportions of individuals in each age group are called age structure of that population. The ratio of the various age groups in a population determines the current reproductive status of thepopulation

Lotha- Voltera model

This model describes the interactions between two species in an ecosystem, a predator and a prey. Since we are considering two species, the model will involve two equations, one which describes how the prey population changes and the second which describes how the predator population changes. The main difference between the Lotka- Volterra model and the single species model is that , this model have two stocks (reservoirs) one for each species. Each species will have its own birth and death rates. In assertion, the Lotka – Volterra model involves four parameters rather than two. When interspecific competition coefficient is less than one it means that the species 2 has inhibiting effect on species 1

Critically endangered species

- Population traits which make a species susceptible to extinction are larger body size, small population size and low reproductive potential, lack of genetic variability, narrow range distribution and island species. A taxon is critically endangered when it is facing an extremely high risk of extinction in the wild in immediate feature. The individuals of the species which have declined to low number is not a genetically open system.
- Reduction of population breeding ability due to increased relatedness through the action of incompatibility mechanisms in plants or behavioural difficulties in animals.
- Loss of some alleles from the species causing loss of genetic diversity with consequent inability to respond rapidly to selection
- Expression of deleterious alleles and increased homozygosity increase mortality of young and inbreeding depression leads to reduced offspring fitness.

www.ingramcontent.com/pod-product-compliance
Lightning Source LLC
Chambersburg PA
CBHW071746240526
45471CB00022B/590